A NATURALIST'S GUIDE TO THE

TREES
OF
BRITAIN
and Northern Europe

Andrew Cleave
Photographic consultant: Paul Sterry

JOHN BEAUFOY PUBLIC

· CONTENTS ·

INTRODUCTION

This book describes the most common and eye-catching trees and large shrubs found in Britain and Ireland, as well as neighbouring areas of mainland Europe, including France (north of the Loire Valley), Belgium, Holland, NW Germany, Denmark, Norway and Sweden.

HABITATS COVERED

The region covered encompasses a huge range of habitats, with different soil types, altitudinal ranges and climates, as a result of which a great variety of native tree and shrub species grow in it. Additionally, many species from other parts of the world have been introduced for commercial forestry, for their fruits and seeds, and for use as ornamental specimens in parks and gardens. Even the most densely populated areas contain trees and shrubs that were planted to soften the harsh lines of buildings, or provide shade and greenery for city dwellers.

Trees and shrubs occur in all of the major habitat types in our region, from coasts and lowlands to the highest regions of the mountains and the Arctic. At very high altitudes only small, shrubby species, such as Woolly Willow, can survive, but lower down the mountain

An urban park planted with flowering trees and shrubs.

Scots Pines Pinus sylvestris form extensive woodlands in many northern and upland areas.

Mixed oak woodland flourishing in a steep-sided valley.

slopes various coniferous species are able to tolerate the cold winters, high rainfall and poor soils. The greatest range of trees and shrubs can be found at lower altitudes. It is here that most broadleaved species and large mixed forests of deciduous trees occur. Agricultural land by necessity has far fewer trees and shrubs, but many species are planted or left to grow naturally on such land to provide shelter belts, stock-proof barriers and hedgerows. Trees that provide food in the form of fruits and nuts are grown in orchards or plantations. They are also planted in smaller numbers in gardens, where they may even remain as isolated specimens long after the adjoining houses have been abandoned. Coastlines can be harsh places for trees and shrubs, where exposure to strong winds and salt spray, and poor, sandy soils or floods are difficulties for many species. However, some species can tolerate these conditions, and provide useful shelter and often help to stabilize thin, sandy soils. Tamarisk and Sea Buckthorn thrive where other species may not be able to grow, and are sometimes planted to provide extra cover on sand dunes and coastal marshes.

Trees are of immense importance to humans because of their many uses. Timber from a variety of trees is used for building homes, and making furniture, tools and utensils, as well as for fuel. Trees that provide important edible fruits and seeds include apples, chestnuts and olives, and many communities rely on firewood for heating and cooking. Trees provide the best renewable source of energy, food, building materials, shelter and habitats for wildlife. With careful management, woods and forests can provide continuous supplies of these essential requirements.

A single mature oak tree can provide a home for a vast range of wildlife, with many species sheltering within it, feeding on its leaves, collecting pollen and nectar, and relying on its annual crop of acorns for a winter food supply. A mixed woodland of native tree species is even more rich in wildlife and is one of our most important natural habitats. Non-native trees, however, are often less hospitable to native invertebrates and are less likely to support much wildlife. They may also become rather invasive and crowd out native species if they are more resistant to pests and diseases.

Many tree species are extremely long lived, and some may be the oldest living organisms on Earth. Yew trees, commonly planted in churchyards, but also occurring naturally in the countryside, are thought to be able to live for well over 1,000 years, and some individuals may be even older than this. Several species of oak are known to live for many hundreds of years, and even smaller trees like hazel, if regularly cut for timber, can regenerate many times over and also live for several hundred years.

Timber floorboards and wooden beams are traditional building materials over a large area of Britain and Europe.

A young English Oak Quercus robur in its first year, having just emerged from the germinating acorn. It is possible that this tiny tree could live for many hundreds of years more, producing immense quantities of acorns during its long life.

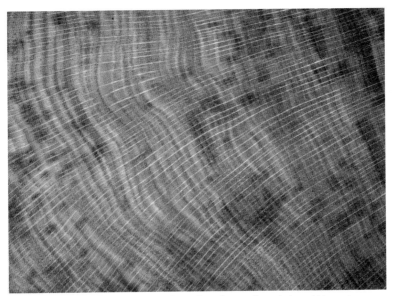

A section through the trunk of an English Oak tree shows the annual growth rings which vary in thickness according to rate of growth in each growing season.

WHAT IS A TREE?

Mature, spreading English Oaks and lofty Noble Firs are obviously trees, but would a straggling Tamarisk or dense Juniper also qualify as trees? Are these smaller species shrubs? There is one feature that is common to both trees and shrubs, and that is that their stems or trunks increase in girth each year by laying down internal layers of woody tissue, each one forming on the outside of the previous year's layer. These layers form the annual rings, or growth rings, which can be seen in the cut section of a tree-trunk. A new ring forms each year so that the number of rings can be used to calculate the age of a tree. In some years the tree may grow well and form a thick ring, and in others growth may be limited so a thinner ring forms.

The stout bole and peeling bark of a London Plane tree, growing on a city street.

Trees are usually defined as having a single main stem of around 5m in height, with a branching crown above this. Shrubs, however, may have many stems arising at ground level, and usually do not reach the height of a tree. Both form woody tissue and growth rings. However, some species may grow as either trees or shrubs, depending on the circumstances. Hazel, for example, can form large trunks and reach a height of several metres, but this very important species is almost always managed by coppicing, where the stems are cut to the ground and allowed to regenerate, producing many smaller stems and a dense, shrub-like appearance. Ash and Birch also respond to coppicing in this way, growing as single-trunk trees in a natural state, but becoming shrubby and multi-stemmed if cut to the ground.

The paired needles of Scots Pine Pinus sylvestris.

A mature cone of Scots Pi

Trees and shrubs belong to a variety of plant families, many of them also including herbaceous species. The rose family, Rosaceae, includes shrubby species like Dog Rose, herbaceous plants like Wild Strawberry, and large trees such as Rowan and Wild Pear.

The plant kingdom is divided into two main classes, the gymnosperms (naked seeds) and angiosperms (hidden seeds). Within the gymnosperms are the Maidenhair Tree, a very primitive species, and the conifers, or cone-

An Acer leaf showing the autumn colour changes from green to red prior to leaf-fall.

bearing trees. Cones are woody structures made up of scales that protect the seeds inside them. Conifers have leaves in the form of needles and are mostly evergreen, retaining the needles throughout the year, but some, like the larches, are deciduous, shedding their needles in autumn. Conifers can live in very harsh environments such as high-altitude mountain slopes, Arctic regions and very dry coastal areas. Some conifers reach a very great size and are among the largest living things on Earth; others are very long-lived, and may in fact be some of the oldest living organisms on Earth.

In angiosperms the seeds are protected inside an ovary, which is a structure that may later develop into a fruit or seedpod. Many of these plants can reach a large size and are an important source of food for numerous species, including humans. Some of these tree species, such as Holly and Holm Oaks, are evergreen, while many more are deciduous, losing their leaves in autumn, often after a spectacular display of colour changes.

IDENTIFYING TREES

Very little equipment is needed to help identify trees, but a notebook, camera and hand lens are useful. Key features to look for are the overall structure and shape of a tree, the texture and colour of the bark, the shape and arrangement of the leaves, and the flowers and fruits. Trees may be described as domed, tall or spreading, but the form can vary

A compound leaf (left) of Rowan and a simple leaf of beech (right)

according to the situation where a tree is growing. A single isolated tree can take on its natural shape, but one growing in a forest is more restricted; and one growing on an exposed hillside, or in very poor soil conditions, may be much more stunted. Leaf shape and arrangement is a very helpful feature. Sometimes it is necessary to look at the leaf stalk, or petiole, to check for hairiness, or the presence of lenticels. Hairs may be present on the underside of a leaf but not the surface. Some leaves grow singly while others are compound, with several arising from a single petiole. The leaf margin may be entire, as in a beech leaf, or toothed as in Hornbeam. Other leaves have a lobed shape with margins that may be entire or toothed.

The bark is an important part of the tree, protecting the living layer of cambium beneath it from damage. Bark may change in texture as a tree grows older, and may also have a covering of algae, mosses and lichens that hide its true colour. Often, however, the epiphytic plants will grow more luxuriantly on one side of the tree than the other, so it is usually possible to see the bark somewhere on the trunk.

Leaf of Hornbeam (left) and Norway Maple (right).

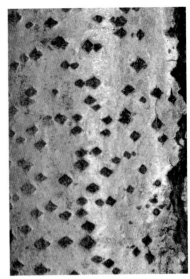

The bark of the Common Pear breaks up into square plates in mature trees, but the bark of Grey Poplar has diamond-shaped fissures set in a silvery-grey smooth bark.

The arrangement of the branches varies greatly between species and with the age of a tree. The oldest branches on the lower part of the trunk may be spreading or even drooping under their own weight, while younger branches higher up are shorter and may be more vertical. Many conifers growing in harsh environments where heavy snowfall is a regular feature have fine needles and drooping branches that allow snow to fall off them rapidly, preventing the branches from breaking. Deciduous trees growing in relatively sheltered lowland environments are more likely to have spreading branches and broader leaves that allow them to trap the maximum amount of the sun's energy. Beech trees have spreading branches and thick leaf cover, and they spread very dense shade on the woodland floor below. Some broadleaved forest trees, like ash, have narrower leaves and more ascending branches, allowing them to rise above other trees nearby. Some deciduous trees growing in exposed situations have very fine, flexible branches that can bend and sway in the wind, and they also have smaller leaves that offer less resistance.

One feature of trees that is usually not possible to study is the root system, but occasionally if a tree blows over the roots may be torn out of the ground. Some trees have a spreading root system that mirrors the arrangement of the branches above, while others have a deeper taproot that grows further down into the soil. The roots have two very important functions. One is to anchor the tree in the ground, and the other is to

The exposed spreading root system of some ancient Beech trees growing on thin chalky soil.

take up the huge quantities of water required by the tree. The main roots become large, strong and woody, like the branches, but these divide into smaller and smaller rootlets, ending in microscopic root hairs that are the structures which take up water and dissolved minerals from the soil. In some trees, such as alders, the roots have many rounded nodules growing on them, and these contain millions of bacteria that help to fix nitrogen from the atmosphere.

The flowers, fruits and seeds of trees and shrubs are a great aid to identification, and seed cases, cones, fruits and nuts can often be found on the ground beneath trees in winter to help identify them. Some trees produce male and female flowers on separate trees, and it may be necessary to find a female tree with fruits to help name a species. Conifers produce separate male and female flowers, sometimes on separate trees and sometimes on different parts of the same branch. The male flowers are usually small and yellow, releasing clouds of dust-like pollen in spring. Female flowers are often in the form of small reddish cones that are slightly sticky so that they can trap the wind-borne pollen and become fertilized. The male flowers wither away soon after they have released the pollen, but the female flowers continue developing, sometimes for as long as three years, often forming the characteristic woody cones that decorate a tree. These open up to release the papery winged seeds, and eventually fall to the ground.

Broadleaved trees have a great range of flower types. Some have male flowers in the form of pendulous catkins, which release clouds of pollen, and female flowers that are inconspicuous and sessile on the ends of twigs. In other species the flowers may be in the form of showy blossom with male and female parts in the same flower. These then give rise to fruits later in the year. Magnolias have some of the largest flowers, but cherries and Almonds, for example, may be covered in smaller flowers, creating a magnificent spectacle in spring.

The winged seeds of Field Maple can be carried far from the tree by the wind.

The times at which leaves open, flowers are produced and fruits form, vary greatly across the region and also from year to year, depending on the weather conditions at the time. In the far south-west of the region spring arrives very early, winters are usually mild and the growing season can be quite long, whereas further north, and at higher altitudes in mountainous areas, spring arrives later, the growing season is shorter, and fruit production and leaf fall may come earlier.

Male and female flowers of Hazel Corylus avellana. The male flowers (left) are pendulous yellow catkins which release pollen when shaken by the wind. The female flowers (right) are small and red with a sticky surface which catches the wind-borne pollen grains.

Ancient woodlands

Most trees can live to a very great age, and there are a few remaining areas of woodland that have had continuous tree cover for hundreds and in some cases thousands of years. There may not actually be vast numbers of ancient trees in these woodlands, because constant regeneration provides a permanent cover of younger trees. However, ancient woodlands do have old trees in them, usually of a variety of species, plus seedlings, immature trees and old, decaying specimens no longer capable of regenerating. All of these afford protection for a wide range of other plants as well as animals that are associated with them. Modern plantations, on the other hand, are more likely to have a much smaller range of species (and sometimes only one tree type), all of the same age, and usually planted in straight lines to facilitate woodland management. Plantations can be valuable habitats, but they are likely to be felled quickly for commercial reasons, thereby rapidly displacing any resident wildlife.

Glossary

acute Sharply pointed.

adpressed Growing very close to a stem.

alien Introduced by humans from another part of the world.

alternate Not opposite (of leaves on stem).

anther Pollen-bearing tip of stamen.

aril Red, fleshy, berry-like structure surrounding seed of a Yew tree.

auricle One of a pair of lobes at the base of a leaf.

axil Angle between upper surface of a leaf or its stalk, and stem on which it is carried.

berry Fleshy, soft-coated fruit containing several seeds.

bole Trunk of a mature tree.

bract Modified, often scale-like leaf found at base of flower stalk in some species.

calcareous Containing calcium, the source typically being chalk or limestone.

capsule Dry fruit that splits to liberate its seeds.

carpel Structure composed of female part of a flower.

catkin Hanging spike of tiny flowers.

clasping Referring to leaf bases that have backwards-pointing lobes which wrap around the stem.

compound Leaf that is divided into a number of leaflets.

cone Woody, seed-bearing structure of pines, larches, spruces, firs and similar trees.

conelet Very small, seed-bearing cone.

coppicing Traditional method of cutting trees and shrubs to the ground to encourage new growth.

cordate Heart shaped at the base.

cultivar Plant variety created by cultivation.

deciduous Plant whose leaves fall in autumn.

decurrent With the base of a leaf extended down the stem.

deflexed Bent sharply downwards.

dehiscent A seed pod which dries out and splits to release seeds.

dioecious Having male and female flowers on separate plants.

distal Situated away from the point of attachment.

entire In the context of a leaf, a margin that is untoothed.

fruits Seeds of a plant and their associated structures.

globose Spherical or globular.

hybrid Plant derived from cross-fertilization of two species.

inflorescence Flowering structure in its entirety, including bracts.

introduced Not native to a region.

involucre Ring of bracts surrounding a flower or flowers.

lanceolate Narrow and lance-shaped leaf.

leaflet Leaf-like segment or lobe of a leaf.

keeled Shaped like keel of a boat.

lenticels Small pores on fruits, stems and twigs.

linear Slender and parallel sided.

lobe Division of a leaf.

midrib Central vein of a leaf.

native Occurring naturally in a region and not known to have been introduced.

naturalized Denotes plant or animal that has been introduced and established as if native.

oblong Leaf whose sides are at least partly parallel sided.

opposite (Usually leaves) arising in opposite pairs on stem.

oval Leaf shape.

ovary Structure containing ovules, or immature seeds.

ovoid Egg shaped.

palmate Leaf with finger-like lobes arising from the same point.

pedicel Stalk of a single flower.

petals Inner segments of a flower, often colourful.

petiole Leaf stalk.

pinnate Leaf division with opposite pairs of leaflets and a terminal one.

pod Elongated fruit, often almost cylindrical, seen in pea family members.

pollen Tiny grains that contain male sex cells, produced by a flower's anthers.

processes Small projections at tips of cone scales.

sessile Without a stalk.

sepal Outer, usually less colourful, segments of a flower.

stamen Male part of a flower, comprising anther and filament.

stellate Star-like.

stigma Receptive surface of female part of a flower, to which pollen adheres.

stomata Small pores in the surface of a leaf.

style Element of female part of flower, sitting on the ovary and supporting the stigma.

whorl Several leaves or branches arising from the same point on a stem.

Maidenhair Tree
■ *Ginkgo biloba* Height >28m

DESCRIPTION Tall, slender deciduous tree with a single trunk and spreading branches bearing greenish-brown side-shoots. Bark is greyish-brown with a corky texture, and becomes deeply ridged in old specimens. Leaves are yellowish-green, >10mm long and fan shaped, with radiating veins reaching to the margins. Mature leaves are divided at least once. Leaves are widely separated on mature shoots, more crowded on young ones. Flowers are rarely seen in our region, but male catkins sometimes form as most trees are male. Fruits are 3cm long, ovoid and green at first, becoming yellow and foul smelling when mature. **HABITAT** Native to China and now endangered in the wild, but widely planted in urban sites, doing best in warmer areas. **DISTRIBUTION** Often encountered in tree collections, parks and city centres, but becoming scarcer in north.

Common Yew ■ *Taxus baccata* Height >25m

DESCRIPTION Broadly conical evergreen with dense, dark green foliage. Trunks of old trees can be gnarled and twisted, with deep red bark peeling to show red patches beneath. Branches are level or ascending, and are usually obscured by dense, needle-like leaves that are arranged in rows on either side of the twig. Each leaf is dark green above and paler below, with two pale yellowish bands. Male flowers are solitary, and formed from clusters

of yellow anthers that release fine, dust-like pollen. Female flowers are solitary, green and after pollination form hard fruits surrounded by a bright red, fleshy aril. A very poisonous tree and the subject of many superstitions. **HABITAT** Native to much of Europe, occurring in dry, lime-rich habitats, and sometimes forming dense woodland. **DISTRIBUTION** Occurs widely in suitable habitats, but also commonly planted in churchyards, parks and urban sites.

California Nutmeg ■ *Torreya californica* Height >20m

DESCRIPTION Broadly conical tree with a stout bole. Reddish-grey bark has narrow vertical ridges. Branches in mature trees are almost horizontal, with descending greenish shoots bearing leaves in rows on each side. Leaves are needle-like with greyish stripes on undersides and a sage-like aroma if crushed. Male trees bear small yellow catkins on undersides of shoots. Female trees produce 5cm-long, ovoid fruits containing inedible seeds that are green with purple streaks and resemble nutmegs. **HABITAT** Native to mountain woodland in California, USA. **DISTRIBUTION** Widely planted in parks, mature gardens and tree collections.

Japanese Cow-tail Pine or Plum Yew ■ *Cephalotaxus harringtonia* Height >6m

DESCRIPTION Small, bushy evergreen resembling Common Yew (see p. 16). Needle-like leaves are densely packed on slightly down-curved twigs. Two variants can be seen in cultivation: var. *drupacea* has shorter leaves with silvery-green undersides growing nearly vertically on curving shoots; var. *fastigiata* is an upright form with darker foliage and leaves >7cm long. Creamy-white male flowers form on undersides of twigs; short-stalked female flowers form on separate trees, giving rise to small, plum-like fruits. **HABITAT** Hillsides and mature gardens. **DISTRIBUTION** No longer seen in the wild in Japan, but widely planted in parks and gardens.

Plum-fruited Yew ▪ *Podocarpus andinus (Prumnopitys andinus)* Height >20m

DESCRIPTION Tall, yew-like tree that may have a single bole and horizontal branches, or several boles and more upright branches. Bark is usually dark grey and smooth. Soft, needle-like leaves are 2.5cm long, and bluish-green above with two pale bands below. Male catkins grow at ends of shoots; female flowers are small and green, growing on separate trees. Fruits resemble small plums, ripening to become black and dusted with a fine bloom.
HABITAT Native to mountain slopes of Chile and Argentina.
DISTRIBUTION Widely planted in parks and gardens, and tolerant of heavy clipping and pruning.

Willow Podocarp

▪ *Podocarpus salignus* Height >20m

DESCRIPTION Large tree with a stout bole and dark orange-brown bark that peels away in strips on mature specimens. May also grow as a smaller, multi-stemmed bush. Willow-like leaves can be >12cm long, and look leathery but feel soft and pliable. Catkin-like male flowers are produced on separate trees from the small female flowers. Small fruits are inedible. **HABITAT** Native to hillsides and mountains in areas of high rainfall in Chile, where it is threatened by habitat loss. **DISTRIBUTION** Widely planted in parks, gardens and urban sites.

Prince Albert's Yew ■ *Saxegothea conspicua* Height >18m

DESCRIPTION Resembles a large Common Yew (see p. 16), with a strong ribbed bole covered with reddish or purple-brown bark, peeling off in rounded scales. Leaves are flattened, curved needles >3cm long, arranged untidily on shoots; they are greenish, have two pale bands on undersides and smell of grass. Male flowers are purplish, and grow in leaf axils on undersides of shoots. Female flowers are small and blue-grey, and give rise to tiny greenish conelets that are borne at tips of shoots. **HABITAT** Native to forests of S Chile and Argentina, preferring damp, sheltered sites. **DISTRIBUTION** Best specimens outside its native area found in Ireland and SW England, but often planted in parks and gardens elsewhere.

Monkey-puzzle ■ *Araucaria araucana* Height >30m

DESCRIPTION Tall evergreen with a cylindrical trunk and greyish, tough, heavily ridged bark bearing numerous rings of old stem scars. Branches are horizontal or slightly drooping, and evenly distributed around trunk. Leaves are 3–5cm long, oval, bright glossy green and scale-like, each with a spiny tip and overlapping the next leaf. Male cones, >10cm long, are borne in clusters at shoot tips. Female cones are rounded, >17cm long, green for the first two years and grow on upper surfaces of shoots; they conceal 4cm-long, edible brown seeds. Trees are either male or female. **HABITAT** Native to mountains of Chile and Argentina; first brought to Europe in 1795. **DISTRIBUTION** Common as ornamental tree in parks and gardens. Grows well in towns, preferring well-drained soils.

Lawson's Cypress ▪ *Chamaecyparis lawsoniana* Height >40m

DESCRIPTION Upright, narrowly conical evergreen with dense foliage. Sometimes grows as a single slender trunk, but often on a repeatedly forked trunk. Bark cracks vertically into long, greyish plates. Trunk bears many small branches, each in turn bearing numerous smaller shoots that are usually flattened and pendulous. Small, scale-like leaves, >2mm long, are flattened along shoot, in opposite pairs, showing paler colours on undersides of shoots. Crushed leaves smell of parsley. Male flowers, borne at tips of twigs, are small reddish cones, >4mm long, and shed pollen in early spring. Female cones, found on the same tree, are >8mm in diameter when young. **HABITAT** Native to mountain slopes of W USA, where trees may reach 60m in height. **DISTRIBUTION** Widely planted in parks and gardens; many colour forms and variants available.

Hinoki Cypress ▪ *Chamaecyparis obtusa* Height >25m

DESCRIPTION Evergreen resembling Lawson's Cypress (see above), with soft reddish bark and mainly level branches. Leaves are blunt pointed, bright green with white lines below, eucalyptus scented and arranged in flat sprays. Rounded female cones are blue-green at first, yellowing with age. Male cones are small and reddish-yellow. Cultivars include a golden form, 'Crippsii', and several dense-foliaged small forms, for example Club-moss Cypress 'Lycopodioides'. Slow-growing dwarf forms are ideal for small gardens. **HABITAT** Native to hillsides and mountains of Japan and Taiwan. Grows best in wet areas, where it may reach a great size. **DISTRIBUTION** Introduced to Britain in 1861. Popular in parks and gardens.

Nootka Cypress
■ *Chamaecyparis nootkatensis* Height >30m

DESCRIPTION Elegant conical evergreen with slightly upturned branches bearing pendulous shoots. Tough, scale-like leaves have an unpleasant smell when crushed. Male flowers are yellow. Female flowers are small cones that are blue in the first year, ripening through green to brown. Cultivars include 'Pendula', with upswept branches and pendulous shoots, and 'Variegata', with golden foliage. **HABITAT** Discovered near Nootka, on Vancouver Island, Canada; occurs elsewhere in Pacific north-west. Hardy, but intolerant of lime-rich soils. **DISTRIBUTION** Often planted in parks and tree collections.

Leyland Cypress
■ *x Cupressocyparis leylandii* Height >35m

DESCRIPTION Evergreen hybrid between Monterey and Nootka Cypresses (see p. 22 and above), first raised in 1888. Normally a tall, narrowly conical tree with reddish-brown bark with thin vertical ridges; this is usually hidden by dense, almost vertical branches and thick foliage. Pointed, scale-like leaves are about 2mm long. Male and female cones are seldom produced, but they occur on the same tree. Male cones are small and yellow, growing at tips of shoots; they release pollen in early spring. Female cones are >3cm across and rounded, with eight scales bearing pointed processes; they are green at first, becoming brown and shiny. **HABITAT** Man-made hybrid, so does not have a native range or habitat, but is tolerant of most soil types. **DISTRIBUTION** Frequently planted in towns and gardens, and often clipped into hedges, suffering mutilations because of proximity to buildings or too vigorous growth.

Monterey Cypress ■ *Cupressus macrocarpa* Height >36m

DESCRIPTION Large evergreen, pyramidal when young, domed and spreading when mature. Reddish-brown bark becomes ridged and scaly with age. Crowded, upright

branches on younger trees become more level and spread with age. Small, scale-like leaves, on stiff, forwards-pointing shoots are lemon scented. Male cones are yellow, >5mm across and produced on tips of shoots behind female cones. Female cones are 2–4cm across, rounded and bright green at first, maturing to purplish-green. Golden-foliaged form 'Lutea' is more spreading and tolerant of sea winds than main species. **HABITAT** Native near Monterey, California, USA, where it is now rare and never attains the size it can in W Britain and Ireland. **DISTRIBUTION** Formerly widely planted in hedgerows and shelter belts.

Italian Cypress

■ *Cupressus sempervirens* Height >22m

DESCRIPTION Slender, upright evergreen with dense dark green foliage. Typically columnar, but sometimes more pyramidal in the wild. Bark is grey-brown and ridged, and branches are upright and crowded, bearing clusters of shoots. Dark green, scale-like leaves are about 1mm long and unscented. Small, greenish-yellow male cones >8mm across grow on tips of side-shoots. Elliptical, yellowish-grey female cones >4cm across grow near ends of shoots; they ripen to brown. **HABITAT** Native to mountain slopes of S Europe and Balkans, east to Iran. **DISTRIBUTION** Elegant columnar form widely planted in parks and formal gardens, especially in south of region.

Smooth Arizona Cypress ▪ *Cupressus glabra* Height >22m

DESCRIPTION In the wild (in Arizona) this species has a spreading habit, but the British form grows into a neat, ovoid tree with blue-grey foliage often with white tips. Reddish or purplish bark falls away in rounded flakes in older specimens, revealing yellow or reddish patches. Small, greyish-green leaves, often with a central white spot, have a grapefruit scent when crushed. Male cones are small, yellow and grow at tips of shoots. Female cones are oval, >2.5cm across when mature and greenish-brown; scales have a central blunt projection. **HABITAT** Native to hillsides and mountains of Arizona, USA. **DISTRIBUTION** Commonly planted in our region for hedging and as an attractive specimen tree.

Patagonian Cypress
▪ *Fitzroya cupressoides* Height >22m

DESCRIPTION Densely foliaged evergreen with reddish-brown bark that peels away in vertical strips. Thick branches grow from low down on the bole and curve upwards to grow almost vertically, bearing descending masses of shoots. Leaves are hard, blunt-ended scales, curving outwards away from shoots, with white stripes on both surfaces. Sometimes a prolific producer of cones, which are small, rounded and brown, and >8mm across. **HABITAT** Native to mountains of Chile and Argentina, sometimes attaining great size and age. Named after Captain Fitzroy of HMS *Beagle*, which carried Darwin on his explorations of South America in the 1830s. **DISTRIBUTION** Hardy and long-lived tree often planted in large parks and tree collections.

Incense Cedar

■ *Calocedrus decurrens* Height >35m

DESCRIPTION Elegant columnar evergreen with a narrowly rounded crown. Bark cracks into large, reddish-brown flakes. Numerous short, upright branches run up trunk from near ground level. Scale-like leaves grow in whorls of four, each bearing a short, incurved, pointed tip, adpressed and concealing shoots, smelling of turpentine when crushed. Male cones are >6mm across, ovoid, deep yellow and borne at tips of lateral shoots. Female cones are 2–3cm across, ovoid and pointed, with six scales; two large fertile scales have outwardly pointed tips. **HABITAT** Native to forests in California and Oregon, USA. **DISTRIBUTION** Very popular ornamental tree in our region; easy to propagate from cuttings and to grow from seed.

Common Juniper

■ *Juniperus communis* Height >6m

DESCRIPTION Usually seen in the form of a small evergreen and aromatic shrub, but sometimes grows into a small, bushy tree, depending on location and climate. Reddish-brown bark peels off in thin strips in older trees. Upright branches bear three-angled, ridged twigs that carry small, pointed, needle-like leaves in whorls of three. Leaves have a pale band on upper surface, and a gin-like aroma when crushed. Male cones are small, yellow and rounded. Female cones are rounded and green, becoming black when mature. **HABITAT** Hillsides, mountain slopes and open woodland, usually on drier soils. **DISTRIBUTION** Extremely widespread in a range of habitats, and often planted in gardens in cultivated forms.

Prickly Juniper or Cade ▪ *Juniperus oxycedrus* Height >14m

DESCRIPTION Spreading
evergreen shrub or small, untidy
tree with brown bark, sometimes
tinged with purple, which peels
away in vertical strips. Sharply
pointed needles are arranged in
whorls of three; upper surface of
each has two pale bands separated
by a slightly raised midrib, and
lower surface has a pronounced
midrib. Female cones are rounded
or pear-shaped, and mature to a
reddish colour. **HABITAT** Native to
S Europe, generally preferring dry,
open habitats. **DISTRIBUTION**
Occasionally planted in parks
and gardens in warmer areas. An
aromatic oil (Cade Oil) is extracted
from the twigs.

Drooping Juniper

▪ *Juniperus recurva* Height >14m

DESCRIPTION Small evergreen
with ascending branches but
drooping foliage, and a broadly
conical outline. Greyish-brown
bark peels in long, untidy shreds.
Tough, needle-like leaves clasp
shoots and have a paint-like smell
when crushed. Male cones are
yellow, and grow in small clusters
at tips of shoots. Female cones
are produced at ends of shoots
and become oval, black and
berry-like when mature, growing
>8mm across. **HABITAT** Native
to hillsides and mountain slopes
of SW China and the Himalayas.
DISTRIBUTION Popular
specimen tree in parks and gardens,
occurring in several cultivars.

Syrian Juniper

■ *Juniperus drupacea* Height >18m

DESCRIPTION Shapely evergreen forming a slender, tall column of compact, bright green foliage. Sometimes the trunk and crown divide to make a more conical tree. Orange-brown bark peels away in thin shreds. Needle-like leaves are >2.5cm long and grow in bunches of three; each leaf is pointed, with a spine and two pale bands on the underside. Male trees produce tiny, bright yellowish-green, oval flowers. Female trees produce tiny green flowers in small clusters at tips of twigs. Flowers develop into rounded, woody cones, about 2cm in diameter, and turning purple-brown when mature. **HABITAT** Native to mountain forests of W Asia; range just extends into Greece. **DISTRIBUTION** Planted in gardens in mild areas, where it makes a fine specimen tree, tolerating a wide range of soils but rarely producing cones.

Chinese Juniper

■ *Juniperus chinensis* Height >18m

DESCRIPTION Large evergreen with dark green foliage and a sparse habit when mature. Reddish-brown bark peels in vertical strips. Level to ascending branches terminate in twigs bearing needle-like leaves, 8mm long, with sharply pointed tips and two bluish stripes on upper surfaces; borne mostly in clusters of three at bases of adult shoots. Mature leaves are small and scale-like, closely adpressed to shoots and smell of cats when crushed. Male cones are small, yellow and grow on tips of shoots. Female cones are rounded, >7mm long and bluish-white at first, ripening to purplish-brown in the second year. Golden-leaved cultivar 'Aurea' (forming a neat column of golden-green foliage) is popular. **HABITAT** Native to hillsides and open areas of Japan and China. **DISTRIBUTION** Often planted in our region in parks, gardens and churchyards.

Meyer's Juniper ■ *Juniperus squamata* 'Meyeri' Height >11m

DESCRIPTION Small, conical evergreen with striking blue-grey foliage when young. Needle-like leaves have a paler stripe on undersides and lie close to stem. Old needles persist on twigs. Bark of mature trees peels in thin, pinkish-brown scales. Small, elliptical cones form on twigs, becoming black when mature. **HABITAT** Native to hills and mountains across the Himalayas. Tolerates poor soils and tough growing conditions, so suited to town gardens. Slow growing and hardy. **DISTRIBUTION** Popular garden tree used for groundcover, shelter and as a specimen tree. May occur in different colour forms.

Phoenician Juniper ■ *Juniperus phoenicia* Height >8m

DESCRIPTION Small evergreen tree or spreading shrub with scaly twigs bearing two types of leaf. Young leaves are each >1.5cm long and 1mm wide, sharply pointed and show pale bands on both surfaces; they grow in bunches of three spreading at right angles. Mature leaves are each only 1mm long, and resemble tiny green scales clasping the twig. Male cones are inconspicuous and borne at ends of shoots. Female cones are >1.4cm long and rounded, and ripen from black, through yellowish-green, to a deep red in the second year. **HABITAT** Native to Mediterranean coasts and Atlantic shores of Portugal. Favours rocky hillsides, woodland margins, coastal sites and cliffs. **DISTRIBUTION** Found in old tree collections in our region.

Western Red Cedar

■ *Thuja plicata* Height >45m

DESCRIPTION Very tall, conical evergreen with a strong, buttressed trunk and upright leading shoot. Reddish-brown bark has fibrous plates. Tiny, scale-like leaves clasp shoots in alternate, opposite pairs; they are glossy, dark green above and paler below with pale markings. Crushed leaves are pineapple scented. Male and female cones form on separate trees. Small yellow or brownish male cones grow at shoot tips. Female cones are ovoid and >1.2cm long, with 8–10 spine-tipped scales. **HABITAT** Native to W USA, forming forests in upland areas and reaching great heights. **DISTRIBUTION** Widely planted for timber, shelter belts and hedging, and as a specimen tree.

Japanese Thuja

■ *Thuja standishii* Height >22m

DESCRIPTION Broadly conical evergreen with reddish-brown bark, peeling in strips or broader flakes. Branches are U shaped with pendent grey-green shoot tips. Tiny, scale-like leaves grow on flattened sprays, and are lemon scented when crushed. Male flowers form at shoot tips and are dark red at first, yellower when open. Female flowers are greenish, and form in separate clusters on tips of different shoots on the same tree; they ripen to red-brown, scaly cones. **HABITAT** Native to Japanese islands of Honshu and Shikoku. **DISTRIBUTION** Commonly planted in our region as an ornamental tree in parks and gardens.

Oriental or Chinese Thuja
▪ *Platycladus (Thuja) orientalis* Height >16m

DESCRIPTION Small, slow-growing tree with foliage in flat vertical sprays; both surfaces are the same shade of green. Tiny, scale-like leaves, lying flat to stems, are unscented. Male flowers are small, yellow-orange and borne on ends of shoots. Female flowers are greenish and become cones with prominent hooked scales. Several cultivars exist, with varying colours of foliage. **HABITAT** Native to China, where it is also widely planted. **DISTRIBUTION** Most often seen as an ornamental tree in parks and gardens. Associated with longevity in China, and wood is used for incense.

Northern White Cedar ▪ *Thuja occidentalis* Height >20m

DESCRIPTION Broadly conical evergreen with orange-brown bark peeling in vertical strips. Leaves form flattened, fern-like sprays of foliage, showing white, waxy bands below. Crushed leaves smell of apple and cloves. Male cones resemble those of Western Red Cedar (see p. 28). Female cones have rounded tips to cone scales. **HABITAT** Native to E North America. Favours swampy areas, riversides and mountain slopes in high rainfall areas. **DISTRIBUTION** Does not thrive our region, except in very wet areas. Various cultivars exist, which are planted in tree collections.

Hiba ▪ *Thujopsis dolabrata* Height >20m

DESCRIPTION Single-boled, conical tree or broad shrub on a divided trunk. Scale-like leaves clasp shoots, and are glossy green above, with white bands below and pointed

curved tips. Shoots form flattened sprays. Small, blackish male cones grow at shoot tips. Rounded female cones form singly on ends of shoots on the same tree. Mature cones are about 1.2cm long and brown. **HABITAT** Native to Japan, preferring wet soils. Often planted in Japan by temples and in formal gardens. **DISTRIBUTION** Planted in parks and gardens, but only thrives in wetter areas of our area as it is not drought tolerant.

Wellingtonia, Giant Sequoia or Giant Redwood
▪ *Sequoiadendron giganteum* Height >50m

DESCRIPTION Outstandingly large evergreen in its native California. Forms a striking, narrowly conical tree with a huge, tapering bole, ridged and fluted at the base, with thick, spongy, rich-red bark. Lower branches, which may not start for several metres above the ground, are pendulous, but upper branches are more level. Scale-like green leaves, >1cm long, clasp shoots; they smell of aniseed when crushed. Small yellow male cones can be abundant at tips of shoots. Female cones are usually solitary, ovoid, and >8cm long and 5cm in diameter when ripe, with a deep brown colour and corky texture. **HABITAT** Native to Sierra Nevada in California, USA, where it grows in groves on western slopes of the mountains. **DISTRIBUTION** Introduced into Europe, where it thrives in wetter western areas, but does not tolerate air pollution.

Coastal Redwood ▪ *Sequoia sempervirens* Height >50m

DESCRIPTION Impressive evergreen growing to be the tallest tree in the world in its native USA. Conical to columnar outline with a tapering trunk arising from a broader, buttressed base. Thick, reddish-brown bark becomes spongy, eventually deeply fissured and peeling. Branches arise horizontally or are slightly pendulous. Leading shoots have scale-like leaves >8mm long, which clasp stem; side shoots have longer, flattened, needle-like leaves >2cm long, lying in two rows. Crushed foliage smells of grapefruit. Small yellow male cones grow on tips of main shoots. Pale brown ovoid female cones, 2cm long, grow singly on tips of shoots. **HABITAT** Native to California and Oregon, USA, growing best in the hills where the permanent sea mists keep the trees supplied with moisture. **DISTRIBUTION** Introduced into Europe, thriving in wetter areas in west and north.

Japanese Red-cedar
▪ *Cryptomeria japonica* Height >35m

DESCRIPTION Tall, narrowly conical evergreen on a rapidly tapering bole with mostly level branches. Thin, hard bark peels in thin shreds. Narrow, claw-like, yellowish-green leaves point towards shoot tips. Foliage character distinguishes this species from Coastal Redwood (see above), as does the thin bark. Male and female cones are produced on the same tree but on different shoots. Small yellow male cones are borne in clusters at tips of shoots. Mature female cones are covered with feathery scales, unique to this tree, giving the appearance of brownish globular flowers. **HABITAT** Native to China and Japan in wetter areas with good soils. Intolerant of cold, dry conditions. **DISTRIBUTION** Commonly planted as an ornamental tree.

Swamp Cypress

■ *Taxodium distichum* Height >35m

DESCRIPTION Deciduous conifer, conical at first, becoming broader and domed with maturity. When growing in or near water, the fluted trunk is surrounded by emergent 'breathing roots'. Pale reddish-brown bark peels in thin, fibrous strips. Branches are upright, but spreading in older trees. Long shoots bear spirally arranged leaves, and alternate side shoots bear flattened leaves set in two ranks. Leaves are >2cm long and pale green; greyish band on undersides has a fine midrib. Mature trees colour well in autumn before shedding needles. Male cones are produced in slender, branching clusters >15cm long at ends of one-year-old shoots. Female cones are globose and woody, and produced on short stalks; they ripen to purplish-brown in the first year. **HABITAT** Native to swampy areas of S and SE USA. **DISTRIBUTION** Introduced to Europe; does grow in relatively dry areas, but does not then produce breathing roots.

Dawn Redwood

■ *Metasequoia glyptostroboides* Height >35m

DESCRIPTION Conical deciduous conifer with shoots and leaves in opposite pairs. Trunk tapers and is buttressed at base, becoming ridged in older trees. Rich reddish-brown bark peels in vertical strips. Leaves are 2.5cm long, flat and needle-like, pale green at first and becoming darker green later; grow on short lateral shoots that are shed in autumn. Male and female flowers are produced on young shoots in separate clusters on the same tree in spring. Males are yellow; females are greenish, producing rounded green, then brown cones about 2.5cm across. **HABITAT** Native to SW China, although unknown as a living tree (known only from fossil records) until 1941. Prefers wet soils. **DISTRIBUTION** Popular subject in parks and large gardens.

Japanese Umbrella Pine ■ *Sciadopitys verticillata* Height >23m

DISTRIBUTION Broadly conical evergreen, often with a finely tapering crown, but may be bushy. Red-brown bark peels in long vertical strips. Needle-like leaves, >12cm long, are borne in umbrella-like clusters. Needles are deeply grooved on both sides, and dark green above but more yellow below. Male flowers are yellow and produced in clusters. Female flowers are green and grow at tips of shoots, ripening into ovoid, 7.5cm-long, red-brown cones after two years. **HABITAT** Native to Japan, preferring sheltered, sunny sites. **DISTRIBUTION** Commonly planted as a specimen tree and some colour variants are available.

Chinese Fir ■ *Cunninghamia lanceolata* Height >25m

DESCRIPTION Broadly conical evergreen conifer with foliage recalling Monkey-puzzle (see p. 19). Reddish-brown bark is ridged in mature trees, and timber is aromatic. Narrow, strap-shaped leaves are pointed, >6cm long and glossy green, with two white bands below. Dead foliage persists inside crown; it looks bright orange in sunlight. Male and female flowers are yellowish, and are produced in clusters at shoot tips. Cones are rounded, scaly, 3–4cm across and green ripening to brown. **HABITAT** Native to China, forming forests in some hilly areas and used as a timber tree. **DISTRIBUTION** Planted in parks and large gardens as an ornamental and specimen tree.

European Silver Fir or Common Silver Fir

■ *Abies alba* Height >47m

DESCRIPTION Fast-growing fir reaching a great size – it held the record for the tallest tree in Britain up to 1960s. White bark on trunk and branches of mature trees, grey on younger trees. Thick needles, >3cm long, notched at tips, grow in two rows on twigs that are covered with pale brown hairs. Erect cones are green at first, maturing to orange-brown, and >20cm long. They eventually disintegrate into fan-like scales and toothed bracts, leaving just the protruding woody axis. Cones normally grow high up. **HABITAT** Native to European mountains. **DISTRIBUTION** At one time widely planted in Britain for timber, but susceptible to aphid attack and vulnerable to late frosts. Tallest trees are mostly found in west of region.

Caucasian Fir

■ *Abies nordmanniana* Height >42m

DESCRIPTION Large, shapely fir with thick foliage. Dull grey bark becomes fissured with age, forming small, square plates. Tough, green, forwards-pointing needles grow in dense rows around brownish twigs; each needle is 1.5–3.5cm long, slightly notched at the tip and grooved above. Male flowers are reddish and grow on undersides of shoots. Female flowers are greener and upright, and borne in separate clusters on the same tree. Cones are found high up on mature trees (30m); they are 12–18cm long, dark brown and resinous, with projecting, down-curved scales. They disintegrate on the tree. **HABITAT** Native to mountainous regions from Turkey eastwards. **DISTRIBUTION** Planted in our area for ornament, and small specimens are popular as Christmas trees.

Korean Fir ■ *Abies koreana* Height >15m

DESCRIPTION Evergreen conifer, usually broadly conical in outline but sometimes dumpy. Branches are level in conical trees. Strap-like, blunt needles are notched at tips and >18mm long; they are dark green above, but whitish on either side of midrib below. Male flowers are yellowish; female flowers are reddish, maturing into bluish-purple cones that ripen to brown and grow upright in clusters on tops of twigs. **HABITAT** Native to mountain slopes in areas of high rainfall in Korea. Very hardy. **DISTRIBUTION** Widely planted in gardens for its attractive shape and conspicuous cones.

Noble Fir ■ *Abies procera* Height >50m

DESCRIPTION Extremely lofty, narrowly conical conifer when mature. Silver-grey or purplish bark develops shallow fissures with age. Youngest twigs are reddish-brown and hairy, with resinous buds at tips. Bluntly pointed needles, 2–3cm long, are grooved on upper surfaces; their blue-grey colour is marked by paler bands on both surfaces. Male flowers are reddish and supported below shoots. Cylindrical female flowers, resembling small cones, are red or green and grow on upper sides of shoots; a green spine emerges beneath each scale. Cones, >25cm long, are held erect on upper sides of branches. They disintegrate in winter, but may be so abundant that branches are damaged by their weight. **HABITAT** Native to Pacific NW USA on mountain slopes in areas of high rainfall. **DISTRIBUTION** Planted in our region since 1850, reaching the greatest size in the north.

Grecian Fir ■ *Abies cephalonica* Height >36m

DESCRIPTION Tall, spreading tree with grey bark showing a hint of orange in young trees, and deeper grey and fissured to form squarish plates in maturity. Rigid, prickly

needles arise from all around hairless red-brown twigs (not in rows); they are >3cm long with two white bands below. Mature cones are upright, rich golden-brown and >16cm long. Down-curved, triangular bracts protrude from between the scales. Mature trees are often heavily loaded with cones. **HABITAT** Native to exposed sites in Greek mountains, reaching high altitudes in places. **DISTRIBUTION** Grows well in dry areas of Britain, but also thrives in wet regions, where it reaches the greatest size.

Giant Fir ■ *Abies grandis* Height >55m

DESCRIPTION Impressive tree when mature, and fast growing, reaching a height of 40m in as many years. Leaves grow in a comb-like arrangement of soft, shining-green needles,

borne in two rows on either side of downy, olive-green twigs. Needles are >5cm long, with a notched tip and two pale bands below; they are orange-scented when crushed. Cones are smooth, <10cm long and produced high up on trees at least 50 years old, breaking up on the tree to release the seeds. **HABITAT** Native to humid upland areas of coastal W USA. **DISTRIBUTION** Planted in our region for ornament and sometimes commercially.

Alpine Fir ■ *Abies lasiocarpa* Height >16m

DESCRIPTION Narrowly conical tree with smooth, greyish-white bark broken up with resinous blisters. Notched needles, >4cm long, are greyish-green above, have two white

bands below and are densely packed on upper sides of shoots, with the central ones pointing forwards. Small male flowers are yellow, tinged red, and grow below shoot. Female flowers are purple and upright, and appear in clusters on the same plant. Cones are cylindrical, >10cm long, and purple, ripening to brown. Variety 'Arizonica' has bluer leaves than species, and corky bark. **HABITAT** Native to high mountain slopes of W USA. **DISTRIBUTION** Planted in our region for ornament.

Spanish or Hedgehog Fir ■ *Abies pinsapo* Height >25m

DESCRIPTION Young trees have a conical shape; they become open crowned and straggly with age. Bark is smooth and dark grey. Bluish-grey, usually blunt needles, >1.5cm long, are

densely arranged all around twigs. Small male flowers are red, opening yellow. Female flowers are green, growing in upright clusters above shoots. Mature cones are cylindrical, tapering, upright and smooth, but with resinous patches. **HABITAT** Native to dry slopes of Sierra Nevada in S Spain, where it is rare. **DISTRIBUTION** Sometimes planted for ornament in mainland Europe where it tolerates calcareous soils, but only thrives in warm climates.

Pacific Silver Fir or Beautiful Fir ▪ *Abies amabilis* Height >32m

DESCRIPTION Large tree with luxuriant foliage, a strong trunk, silvery bark and a thick, tapering crown when grown in suitable wet climates. Needles are 3cm long, shiny, silvery below, densely packed and orange-scented when crushed. Smooth oval cones are tinged purple and grow on upper surfaces of twigs. **HABITAT** Native to wetter, mountainous areas of NW USA. **DISTRIBUTION** Widely planted in cooler parts of Europe for ornament.

Deodar
▪ *Cedrus deodara* Height >36m

DESCRIPTION Broadly conical evergreen with drooping leading shoot on the tapering crown and drooping tips on branches. Bark is dark on old trees, fissured into small plates. Dark green needles, with pale stripes, grow in whorls of 15–20 on short lateral shoots, or in spirals on larger twigs. Male flowers are purplish, turning yellow with autumn pollen release. Female cones are solid, barrel shaped, >14cm long and 8cm across, and grow only on older trees. **HABITAT** Native to mountain slopes of W Himalayas. **DISTRIBUTION** Introduced into Britain in 1831, and widely planted in parks and gardens, where it can form stately trees.

Atlas Cedar ■ *Cedrus atlantica* Height >40m

DESCRIPTION Broadly conical or
pyramidal tree, with the leading shoot often
rising above the domed top in old trees.
Dark grey bark cracks into large plates with
deep fissures. Tips of branches and shoots
are angled upwards. Shiny, deep green
needles, 1–3cm long, grow in clusters. Male
cones are 3–5cm long and pinkish-yellow.
Ripe female cones are squat, about 8 x
5cm, with a sunken tip and small central
boss. Most frequent cultivar is **Blue Atlas
Cedar** *C. atlantica* var. *glauca*, with bright
bluish-grey foliage; it is hardy and tolerates
atmospheric pollution. **HABITAT** Native to
humid slopes of Atlas Mountains in North
Africa. **DISTRIBUTION** Widely planted for
ornament.

Cedar of Lebanon
■ *Cedrus libani* Height >40m

DESCRIPTION Mature tree is flat topped with an
immense trunk, and dark grey, fissured bark that
becomes dark brown and ridged in very old trees.
Main branches are massive and ascending; smaller,
lateral branches are level, supporting flat plates of
foliage. Needles are >3cm long, and usually grow
in clusters of only 10–15 on short shoots, or singly
if growing on long shoots. Male cones are greyish
or blue-green, and erect, >7.5cm long. Female
cones are solid, ovoid, >12cm long and 7cm across,
ripening from purple-green to brown. Similar
Cyprus Cedar *C. brevifolia* (height >21m) has dark
green needles shorter than those of other cedars
(2cm), and a more open crown. Female cones, >7cm
long, ripen from purple-green to brown. **HABITAT**
Native to mountain forests of E Mediterranean,
where it is scarce. Cyprus Cedar is native to Troodos
Mountains, Cyprus. **DISTRIBUTION** Widely
planted in parks and gardens since 1640, often
reaching great size. Cyprus Cedar sometimes grown
in collections in N Europe.

Common Larch
■ *Larix decidua* Height >35m

DESCRIPTION Deciduous conifer forming a tall, narrowly conical tree, but more often seen in close rows in plantations. Rough, greyish-brown bark in young trees becomes fissured with age. Branches are mostly horizontal, but lower ones on old trees are slightly drooping. Needles grow in tight bunches of >40, each needle >3cm long, and fresh green when first open, becoming darker; each needle has two pale bands below in summer, changing through red to yellow before falling in autumn. Male flowers are small, soft yellow cones. Female cones are red in spring, and persist on twigs after releasing their seeds. **HABITAT** Native to mountains of central and E Europe, tolerating a wide range of soils and climates. **DISTRIBUTION** Widely planted as a valuable timber tree or for ornament.

Japanese Larch
■ *Larix kaempferi* Height >40m

DESCRIPTION Deciduous conifer resembling Common Larch (see above), but lacking the drooping shoots, and having a more twiggy appearance with a dense crown. Reddish-brown bark flakes off in scales. Needles grow in tufts of about 40 and are slightly broader and greyer in colour than those of Common Larch. Male cones are similar to those of Common Larch, but female cones are pink or cream in spring, becoming brown and woody in autumn, and differing from those of Common Larch in having turned-out tips to the scales, so they look like woody rose-buds. **HABITAT** Native to hills and mountains of Japan. **DISTRIBUTION** Commonly planted in Europe for commercial forestry because of rapid growth rate.

Dahurian Larch ■ *Larix gmelinii* Height >30m

DISTRIBUTION Slender, conical deciduous tree with reddish-brown, scaly bark. The mostly level branches sometimes form flattish areas of foliage, and support long, yellowish or red-brown, downy shoots.

Blunt-tipped needles are bright green above with two paler bands below; grow in clusters of 25. Female cones are similar to those of other larches, with pinkish or greenish, slightly projecting bracts that become brown when ripe, with square-ended scales. The variant Prince Rupert's Larch has larger cones. **HABITAT** Native to hillsides and mountains of E Asia. **DISTRIBUTION** Planted in Europe for timber or as a specimen tree. Prince Rupert's Larch is the most widely planted variant.

Western Larch ■ *Larix occidentalis* Height >30m

DESCRIPTION Largest of all the larches, although rarely reaches its maximum height away from its native range. Tall, slender, conical tree with grey, scaly bark forming deep fissures low down. Slightly ascending branches have red-brown shoots and soft needles,

>4cm long, borne in tufts on side shoots. Male flowers are yellow, growing below shoots; female flowers are red and upright above shoots on the same tree. Both open in spring. Cones are ovoid, 4cm long, with long bracts protruding from between the scales, separating this species from all other larches. **HABITAT** Native to mountains of British Columbia, Canada, south to Oregon, USA, in areas of high rainfall. **DISTRIBUTION** Introduced into Europe in 1880s; there are now some very fine specimens in mature European collections.

Norway Spruce
▪ *Picea abies* Height >44m

DESCRIPTION Familiar evergreen, commonly used when small as a Christmas tree. Narrowly conical with a slender, unbranching trunk. Brownish, scaly bark with resinous patches on older trees. Almost level branches carry stiff, short, four-angled needles on short pegs, spreading to expose under-surfaces of twigs. Male cones are small, yellowish and clustered near tips of shoots. Female cones are >18cm long, narrowly oval and pendulous. **HABITAT** Native to European mountains, and lower altitudes further north. **DISTRIBUTION** Widely planted outside its native range, especially for Christmas trees and shelter belts, and occasionally as specimen trees.

Blue Colorado Spruce
▪ *Picea pungens* Height >30m

DESCRIPTION Slender, conical evergreen with purplish, ridged bark. Branches carry smooth, yellowish-brown twigs that support sharply pointed, stiff needles >3cm long and usually dark green (some cultivars are markedly blue-green). Needles grow all around shoots; upper surfaces of shoots have more needles, and some curve upwards to make top surfaces look more dense. Male and female flowers grow in small, separate clusters on the same tree; male cones are red tinged, female cones greener. Mature female cones are pendent, narrowly oval, slightly curved and >12cm long; scales have irregularly toothed tips. Most commonly seen cultivar is var. *glauca* (height >23m), favoured for its attractive bluish foliage. **HABITAT** Native to dry, stony mountain slopes and stream-sides of SW USA. **DISTRIBUTION** Widely planted for ornament and timber throughout much of N Europe.

Brewer's or Weeping Spruce
■ *Picea breweriana* Height >20m

DESCRIPTION Distinctly conical evergreen with a slender bole and dark grey-purple, scaly bark. Branches bear pale brownish or pink, downy twigs with a striking 'weeping' habit of shoots along branches. Flattened needles are sharply pointed, >3cm long, green above with white bands below, and grow all around shoots, often curving forwards. Male flowers are large for a spruce, >2cm across, and reddish. Female cones are pendent, cylindrical, >12cm long, starting purplish but ripening to brown. Overlapping scales have blunt, rounded tips. **HABITAT** Native to mountains and hillsides of W USA. **DISTRIBUTION** Frequently planted in our region as an ornamental, and if given space becomes very graceful in maturity.

Oriental Spruce
■ *Picea orientalis* Height >40m

DESCRIPTION Dense-foliage evergreen that grows into a strongly conical tree with a short, stout bole and pale brown, scaly bark. Slender branches bear numerous hairy twigs and very short, blunt needles, >1cm long, and square in cross-section. Small male flowers are red, then yellow. Female cones, >8cm long, are pendent and ovoid, often curved and green with purple or grey tinges when still growing, ripening to shiny brown. Variety 'Aurea' has bright yellow young foliage. **HABITAT** Native to mountain forests of Caucasus and NE Turkey. **DISTRIBUTION** Widely planted in Europe for ornament, and occasionally for commercial forestry. Variety 'Aurea' is a popular arboretum tree.

White Spruce
■ *Picea glauca* Height >24m

DESCRIPTION Narrowly conical evergreen, broadening out when mature. Purple-grey bark has roughly circular scales. Branches turn upwards at tips, and bear hairless greyish twigs and blunt buds. Pointed needles are four angled, >1.3 cm long and pale green (sometimes bluish), with an unpleasant smell when crushed. Female cones are about 6cm long and 2cm across, cylindrical, pendent and orange-brown when ripe, with rounded margins to scales. **HABITAT** Native to hills and mountains in N North America. **DISTRIBUTION** Widely planted in Europe for timber, shelter belts and ornament.

Black Spruce
■ *Picea mariana* Height >19m

DESCRIPTION Slender, conical evergreen with grey-brown, scaly bark and the shortest needles and cones of any spruce, apart from Oriental Spruce (see p. 43), whose needles are darker green and blunt. Bluntly pointed needles are blue-green above and pale blue below, >1.5cm long, four angled, and grow all around hairy, yellowish shoots. Cones are ovoid, reddish and pendent, >4cm long, and usually grow near tree top. **HABITAT** Native to boreal forests and taiga in North America, often growing on margins of swamps and marshes. **DISTRIBUTION** Planted in Europe as an ornamental tree because of its relatively small size and neat shape.

Sitka Spruce

■ *Picea sitchensis* Height >52m

DESCRIPTION Large, conical evergreen
tapering to a spire-like crown. Trunk is stout
and buttressed in large specimens, with greyish-
brown bark that becomes purplish and scaly.
Ascending branches bear slightly pendent,
hairless side shoots. Needles are >3cm long, stiff
and flattened with a distinct keel; bright green
above with two pale blue bands below. General
impression is of tough, sharply spined, blue-
green foliage on a sturdy tree. Female cones
are yellowish and small at first, growing >9cm;
they become cylindrical and shiny pale brown,
covered with papery toothed scales. **HABITAT**
Native to high-rainfall areas on W coast of
North America, where some specimens that
are protected in national parks have reached
heights of 80m. **DISTRIBUTION** Introduced to
our region and widely planted for commercial
forestry and sometimes for ornament.

Serbian Spruce ■ *Picea omorika* Height >30m

DESCRIPTION Narrowly conical to columnar tree, with a slender form unlike that of all
other spruces. Orange-brown bark becomes scaly in older trees. Lower branches are short
and slightly descending, with raised tips;
higher branches are mostly level or ascending.
Flattened and keeled needles, >2cm long, may
be blunt or barely pointed, and are dark blue-
green above with two pale bands below. Male
cones are large and red, becoming yellow
when releasing pollen. Female cones grow
on curving stalks and are >6cm long, ovoid
and blue-green at first, ripening to brown.
Cone scales are rounded with finely toothed
margins. **HABITAT** Native to limestone rocks
of Drina Basin in Serbia; unknown until 1875.
DISTRIBUTION Popular ornamental tree
because of its pleasing shape, tolerance to a
wide range of soil types, ability to grow in
polluted air near towns and resistance to frost
damage.

Western Hemlock-spruce ■ *Tsuga heterophylla* Height >45m

DESCRIPTION Large, narrowly conical evergreen with dense foliage and a spire-like crown with drooping leading shoot. Needles are dark glossy green above with two pale bands below; they grow in two flattened rows on either side of shoots. Scientific name derives from two leaf sizes – some leaves are 6mm long, others >2cm long, both with rounded tips and toothed margins. Male flowers are reddish at first, but yellow with pollen. Female cones are solitary, ovoid and pendent, >3cm long; scales are blunt. **HABITAT** Native to high rainfall areas of NW USA and Canada. **DISTRIBUTION** Widely planted in Britain and mainland Europe, and reaches a great size. Grows well on most soils except chalk.

Eastern Hemlock-spruce
■ *Tsuga canadensis* Height >30m

DESCRIPTION Untidy-looking tree with heavy branches, a forked trunk, blackish bark and dark foliage. Needles are more tapering than those of Western Hemlock-spruce (see above), with narrower tips; a further row of leaves along middle of shoot twists to show white undersides. Male flowers are small, yellowish and clustered along undersides of shoots. Female cones are 1.5cm long; cone scales have thickened edges. **HABITAT** Native to E North America, preferring areas of high rainfall, and growing in rocky areas. **DISTRIBTION** Widely planted in Britain and mainland Europe for commercial forestry, and also used for paper manufacture.

Douglas Fir ■ *Pseudotsuga menziesii* Height >60m

DESCRIPTION Tall, slender, conical evergreen. Greyish-green, smooth bark with resinous blisters in young trees turns red-brown with age. Whorls of branches support pendulous masses of dense aromatic foliage. Needles are >3.5cm long, blunt or slightly pointed, dark green and grooved above, with two white bands below. Male flowers are small, yellow and pendulous, growing near tips of twigs. Female flowers, resembling tiny, pinkish shaving brushes, grow at tips of twigs. Mature cones are ovoid with unique three-tailed bracts between scales. Named after the Scottish explorer David Douglas. **HABITAT** Native to temperate rainforest of W North America, where it reaches an immense size. **DISTRIBUTION** Planted in Britain and mainland Europe for timber and ornament, doing well in north of region.

Scots Pine ■ *Pinus sylvestris* Height >36m

DESCRIPTION Conical evergreen when young, but more open and flat topped with a long bole when older. Reddish- or grey-brown, scaly bark low down on trunk, but markedly red or orange and papery higher up in mature trees. Broken-off stumps of old branches remain on trunk lower down. Needles, borne in bunches of two, are grey-green or blue-green, each >7cm long, and usually twisted with a short point at the tip. Male flowers are yellow and borne in clusters at ends of previous year's shoots. Female flowers grow at tips of new shoots; usually solitary and crimson at first, becoming brown and persisting in winter. In the second summer they enlarge, becoming green and bluntly conical, and ripening to grey-brown by autumn; they do not open their scales and shed seeds until the following spring. **HABITAT** Native to Scotland and originally much of Britain, and mainland Europe from Spain to Siberia and Turkey. Favours mountain slopes, rocky areas and lowlands on poor soils. **DISTRIBUTION** Widely planted for commercial forestry.

Beach and Lodgepole Pines ■ *Pinus contorta* and *Pinus latifolia*
Height >30m

DESCRIPTION Small to medium-sized evergreen that occurs as two subspecies. Blackish-brown bark occurs in all trees. Scientific name of Beach Pine (ssp. *contorta*) was gained from the twisted appearance of the branches contorted by wind. Needles are paired, >7cm long, with sharp points, and are usually twisted and densely packed on young shoots, but sparser on older shoots. Lodgepole Pine (ssp. *latifolia*) is similar, but more columnar with

a less dense crown; it usually grows on a much straighter, sometimes divided trunk. Needles are broader than those of Beach Pine, and more spread apart. Male flowers grow in dense clusters near tips of shoots. Female flowers grow in groups of up to four near tips of shoots. Cones are rounded to ovoid, >6cm long and 3cm in diameter, and usually shiny yellow-brown. Each cone scale has a slender, sharp tip, which easily breaks off. **HABITAT** Native to coastal W North America, growing in lowland coastal sites and upland areas. **DISTRIBUTION** Widely planted in Britain and mainland Europe for timber on poor soils and exposed, often upland sites.

Bosnian Pine
■ *Pinus heldreichii (leucodermis)*
Height >30m

DESCRIPTION Broadly pyramidal tree with a tapering bole and grey bark with irregular plates. Whitish patches appear with age. Paired needles, >9cm long and densely packed on shoots, are stiff, project at right angles and are pungent when crushed. Cones are >8cm long and 2.5cm across, narrowly ovoid and ripen to brown; each scale has a recurved prickle. Second-year cones are deep blue. **HABITAT** Native to Balkans and SW Italy, mainly on dry mountain limestone. **DISTRIBUTION** Planted in Britain and mainland Europe for ornament, thriving on free-draining soils in drier areas.

Austrian Pine ■ *Pinus nigra* ssp. *nigra* Height >30m

DESCRIPTION Broadly conical, heavily branched tree, usually with a single bole and

narrow crown; greyish-brown bark becomes rough in older trees. Paired needles, >15cm long, are flattened and stiff with finely toothed margins; they last for up to four years, creating dense foliage. Mature cones, >8cm long, have keeled, spined scales; they are usually solitary or grow in small clusters. **HABITAT** Native to mountains of central Europe, and rather variable in appearance across its range. **DISTRIBUTION** Planted widely for shelter and ornament, tolerating a range of soils and climates.

Corsican Pine

■ *Pinus nigra* ssp. *maritima*
Height >30m

DESCRIPTION Similar tree to Austrian Pine ssp. *nigra* (see above), but more shapely when mature. Branches are shorter than in Austrian Pine, and level, so young trees are columnar. Soft, narrow needles grow in pairs, are paler green than those of ssp. *nigra*, >15cm long and often twisted in young trees. Cones are very similar to those of ssp. *nigra*. **HABITAT** Native to Corsica, S Italy and Sicily, growing on hillsides, coastal cliffs and mountain slopes. **DISTRIBUTION** Planted outside its native region on lowland heaths and coastal dunes, and in poor soils. Resistant to pollution.

Mountain Pine
■ *Pinus mugo* Height >30m

DESCRIPTION Occurs in two forms:
the tree-sized ssp. *uncinata* and shrub-like
ssp. *mugo* of high altitudes. Greyish-black
bark is seen in both forms. Bright green
needles, in all trees, are >8cm long,
curved, stiff and appear whorled. All trees
have male flowers in clusters near shoot
tips. Female flowers are reddish, borne in
groups of 1–3. Ripe cones are ovoid, pale
brown and >5cm long; scales each have a
small prickle. HABITAT Native to Alps,
Pyrenees and Balkans; dwarf forms occur
at high altitudes. DISTRIBUTION Tree-
sized subspecies is planted for forestry,
shelter, stabilizing sand dunes and other
such uses. Dwarf forms are favoured
for ornament and are suited to smaller
gardens.

Stone Pine
■ *Pinus pinea* Height >30m

DESCRIPTION Broad, umbrella-shaped
tree with a dense mass of foliage on
spreading branches on top of a tall bole.
Reddish-grey fissured bark, on old trees
flaking away to leave deep orange patches.
Paired needles, >20cm long and 2mm wide,
are slightly twisted; through a hand lens
12 lines of stomata can be seen on outer
surface and six on inner surface. Cones are
rounded to ovoid, >14cm long and 10cm
across, and ripen to rich glossy brown after
three years. Closely packed scales with a
slightly pyramidal surface conceal large,
slightly winged seeds. HABITAT Native to
Mediterranean coasts on sandy and limestone
soils. DISTRIBUTION Occasionally planted
outside its native region, usually near coasts.
Large, edible seeds are harvested and sold as
pine kernels, or pine nuts.

Maritime Pine

▪ *Pinus pinaster* Height >32m

DESCRIPTION Large conifer with a sturdy, slightly tapering bole, often curved in exposed coastal areas; crown is fairly open, reflecting curve of bole. Yellowish-brown bark breaks up into rectangular flakes. Needles are longer and thicker than those of any other two-needle pine. Male flowers are yellow and ovoid, growing in clusters near shoot tips. Female cones are ovoid, red at first and grow in small clusters; they ripen to become conical and woody with a greenish-brown gloss. **HABITAT** Rocky coastal sites along SW Atlantic coasts of Europe and Mediterranean. Grows well on poor, sandy soils, often on heaths and near coasts. **DISTRIBUTION** Widely planted for forestry and shelter belts in warm parts of Europe.

Aleppo Pine

▪ *Pinus halepensis* Height >20m

DESCRIPTION Small pine often growing in a gnarled and deformed manner, but sometimes maturing to form a broad, shapeless crown on a stout bole. Young trees have shiny, smooth, silvery-grey bark, becoming scaly, fissured and redder with age. Twigs are characteristically very pale grey. Paired needles are slender (0.7mm), >15cm long, sometimes slightly twisted and have very finely toothed margins. Red-brown cones are >12cm long and 4cm across, oval or conical, and borne singly on short stalks, or in groups of 2–3, and sometimes deflexed. Cone scales are shiny reddish-brown, with winged seeds >2cm long. **HABITAT** Dry, stony or sandy areas in Mediterranean region. **DISTRIBUTION** Planted elsewhere in Europe for ornament, and occasionally forestry and shelter belts.

Monterey Pine ■ *Pinus radiata* Height >45m

DESCRIPTION Large, variable pine, slender and conical when young, and becoming domed and flat-topped on a long bole with age. Fissured grey bark blackens with age. Main

branches sometimes hang low enough to touch the ground. Bright green needles in bunches of three are thin and straight, >15cm long, each with a finely toothed margin and sharp-pointed tip. Male flowers grow in dense clusters near ends of twigs. Female cones grow in clusters of 3–5 around tips of shoots, ripening to large, solid woody cones, >15cm long, with an asymmetrical shape. Cone scales are thick and woody with rounded outer edges, concealing black, winged seeds. **HABITAT** Native to small coastal area around Monterey, California, USA, and Guadalupe Island and Baja California, Mexico. **DISTRIBUTION** Widely planted in mild areas of Europe as a shelter-belt tree and for ornament, growing well next to the sea.

Northern Pitch Pine
■ *Pinus rigida* Height >20m

DESCRIPTION Narrowly conical tree with stiff, tough needles in clusters of three, and small cylindrical or rounded cones with thinner, but stiff (hence *rigida*) scales. A striking feature, unique to this pine species, is sprouting foliage on the bole that can sometimes lead to multi-stemmed growth forms. Used as a source of pitch for shipbuilding in the past. **HABITAT** Native to east coasts of North America. Tolerant of poor conditions, sometimes growing in lowland swampy areas. **DISTRIBUTION** Occasionally planted in our area for forestry on sites where other trees would not flourish, mainly to provide pulp wood.

Ponderosa Pine or Western Yellow Pine

■ *Pinus ponderosa* Height >40m

DESCRIPTION Large, slender, conical pine with a sturdy straight bole and scaly, pinkish-brown bark. Needles grow in threes, and are >30cm long, narrow (3mm), and stiffly curved, with finely toothed edges and sharp, pointed tips; they are clustered densely on shoots and persist for three years. Cones are ovoid, >15cm long and 5cm across, and form on short stalks or directly on twigs, sometimes leaving a few scales behind when they fall. Cone scales are oblong with swollen, exposed, ridged tips hiding 5cm-long, oval, winged seeds. **HABITAT** Native to mountainous areas and river valleys of W USA. **DISTRIBUTION** Planted in Europe for forestry and ornament, and noted for its very heavy timber.

Arolla Pine ■ *Pinus cembra* Height >29m

DESCRIPTION Densely crowned conifer with a strong bole and slender-conical shape when mature. Reddish-grey bark is peeling and resinous. Needles grow in crowded bunches of five, >8cm long, and can be almost erect. Leaf margins are faintly toothed. Squat cones, >8cm long, grow on short stalks; they are violet-blue first, and ripen to rich brown. Scales are rounded; cones usually fall intact. **HABITAT** Native to high mountain slopes of Alps and Carpathians. **DISTRIBUTION** Widely planted for ornament and shelter. Very hardy in cold climates, and disease resistant.

Bhutan Pine

■ *Pinus wallichiana* Height >35m

DESCRIPTION Narrowly columnar pine, becoming more shapeless with age. Greyish-brown bark is quite resinous. Lower branches are spreading, upper ones ascending. Needles occur in fives, and are >20cm long and 7mm wide; they are supple, with finely toothed margin. Cones are long, cylindrical, >25cm long, grow below the shoot, and are light brown and resinous. Cone scales are wedge shaped, grooved and thickened at the tip. Basal scales are sometimes reflexed. **HABITAT** Native to high mountains and steep valleys of the Himalayas, preferring high-rainfall areas. **DISTRIBUTION** Planted in Europe as an ornamental or specimen tree.

Weymouth or White Pine ■ *Pinus strobus* Height >32m

DESCRIPTION Mature tree has a tapering trunk with dark grey bark, level branches and a rounded crown. Tuft of hairs grows below each 10cm-long bunch of five blue-green needles, which are protected by a deciduous sheath. Slender cones >16cm long grow beneath twigs; basal scales often curve outwards. **HABITAT** Native to North America, often growing on poor soils and in very wet areas, and forming dense stands in places. **DISTRIBUTION** Planted in Europe mainly for its valuable timber; especially good for carving.

Macedonian Pine
▪ *Pinus peuce* Height >30m

DESCRIPTION Large, narrowly conical pine with a slender trunk, greyish-green bark and a pointed crown. Slender, supple needles, in bunches of five, are >12cm long, with toothed margins and pointed tips. Cones are >20cm long, mostly cylindrical, sometimes curved near the tips and grow below shoots; they are green at first, ripening to brown, and have long-winged seeds. **HABITAT** Native to mountains of Balkans, reaching high altitudes, and very tolerant of low temperatures. **DISTRIBUTION** Planted outside its native area in Europe, especially in colder regions, where it is very hardy and disease resistant.

Rocky Mountain Bristlecone Pine ▪ *Pinus aristata* Height >10m

DESCRIPTION Small, slow-growing pine with a stout bole and grey-brown, scaly bark. Needles grow in bunches of five, and are 2–4cm long, dark green and often flecked with white resin; they are turpentine scented and persist for many years. Cones are >6cm long and have a 6mm-long spine on each scale. **HABITAT** Native to high altitudes of Rocky Mountains, USA, where some trees reach a very great age. **DISTRIBUTION** Occasionally planted in Europe in tree collections and gardens. Tolerant of very harsh conditions.

Bay Willow
■ *Salix pentandra* Height >18m

DESCRIPTION Broadly domed, open-crowned tree when growing in its typical stream-side habitat, but more slender and upright in woodland. Shoots are olive-green and glossy, and leaves are glossy green, showing a bluish tint below. In contrast to other willows, the bright yellow, upright male catkins appear at the same time as the new growth of leaves, rather than just before the leaves. Female catkins are dull yellowish-green, longer and more pendulous. HABITAT Native to Europe in moorland, stream-sides and boggy areas, and also damp upland woods. DISTRIBUTION Widespread and relatively common in places, but only very occasionally planted as an ornamental tree.

Crack-willow
■ *Salix fragilis* Height >25m

DESCRIPTION Large tree when mature, with a broadly domed crown and thick bole with dull grey-brown bark covered with interlocking criss-crossed ridges. Branches arise from near the base. Dull reddish-brown shoots become brighter in early spring. Leaves are long and glossy, on short petioles with toothed margins, widely spaced on shoots. Lower surface is slightly paler than upper surface. Male flowers are yellow, pendulous catkins, opening in early spring. Female catkins are green and pendulous, growing on separate trees. Twigs snap easily and cleanly (hence *fragilis*), and root very readily if stuck in the ground. Often, lines of trees of the same sex grow along riverbanks, all derived from pieces of the same tree. HABITAT Very widespread native species, found in damp lowland woodland and along river and canal banks. Hybridizes freely and is not always true to the type species. DISTRIBUTION Widespread across region.

White Willow
■ *Salix alba* Height >25m

DESCRIPTION Large, broadly columnar tree with a dense crown when growing in the open, but more slender when crowded. Bark is dark grey and shoots are yellowish-grey, downy at first. Bluish-grey leaves are smaller than Crack-willow's (see p. 56). Male catkins are small, elongated-ovoid and pendulous. Female catkins are longer, slender and green. **HABITAT** Widespread native tree, often growing in damp lowland habitats. **DISTRIBUTION** Common in suitable habitats. Easily raised from cuttings and grows well in damp soil, especially near ponds and rivers. So-called Coral-bark Willow cultivars are popular in parks and gardens.

European Violet-willow
■ *Salix daphnoides* Height >12m

DESCRIPTION Domed shrub or small tree with grey bark with shallow ridges. Violet-blue shoots have a light bloom – the best identification feature for this species. Leaves are narrowly ovate to oblong, dark green and shiny above, blue-grey below, with finely toothed margins. Catkins, appearing in early spring before the leaves, are distinctive with their attractive black flecks. **HABITAT** Native to mainland N Europe, growing in upland river valleys and on mountain slopes. **DISTRIBUTION** Introduced to UK and planted for ornament (for attractive colour of twigs), and naturalized occasionally in damp ground. When coppiced produces an abundance of striking slender twigs, which are used for basket weaving.

Siberian Violet-willow
■ *Salix acutifolia* Height >10m

DESCRIPTION Sometimes also known as Violet Willow. Similar to European Violet-willow (see p. 57), but with subtly different shoots and leaves. Shoots are very slender, drooping at the tips and with a waxy feel; they turn violet in winter, with a white bloom. Leaves are longer than those of European (>16cm), narrower and shiny green on both sides. Male catkins are silvery-white with gold anthers; female catkins are green and about 3cm long. **HABITAT** Native to Russia. Grows in marshy ground beside rivers and lakes, and wet woodland clearings. **DISTRIBUTION** Sometimes planted in our area by stream-sides and in large gardens.

Corkscrew Willow
■ *Salix matsudana (tortuosa)* Height >18m

DESCRIPTION Distinctive willow, recognized by its contorted young stems and pointed, twisted leaves. Older shoots are less twisted, but bark still shows signs of earlier curves, and even boles of older trees show some torsion. Leaves open in early spring and are bright green, darkening by summer. Male flowers are yellow catkins, >2cm long. Female flowers are smaller, greenish catkins, borne on separate plants. Both open at around the same time as the leaves. **HABITAT** Originally from China, but now found only in cultivation. **DISTRIBUTION** Thrives in a range of habitats, including parks and gardens. Increasingly popular; grows rapidly, even in dry soils.

Weeping Willow
■ *Salix* x *sepulcralis* var. 'Chrysocoma'
Height >15m

DESCRIPTION Hybrid between **Chinese Weeping Willow** (*S. babylonica*) and White Willow (see p. 57), with features from both. Bark is greyish-brown and deeply fissured. Very pendulous branches, and slender golden shoots and foliage, look elegant in waterside settings. Catkins appear in spring with the leaves, and both sexes grow on the same tree and sometimes merge in the same catkins.
HABITAT Frequently planted in urban waterside settings. **DISTRIBUTION** Widespread in Europe and a very hardy tree that survives well in colder regions.

Common Osier
■ *Salix viminalis* Height >6m

DESCRIPTION Spreading shrub or small tree. Rarely reaches full potential, being regularly cropped for its long, flexible twigs (withies), which are used for weaving. Natural crown is narrow with slightly pendulous branches. Straight twigs are flexible, covered with greyish hairs when young, and becoming smoother and shiny olive-brown with age. Narrow, tapering leaves are >15cm long, with the margin usually waved and rolled under; leaf underside has grey woolly hairs. Male and female catkins, >3cm long, appear before leaves on separate trees, and are erect or slightly curved. Male catkins are yellow, female browner.
HABITAT Common native tree in wet habitats, especially along riverbanks.
DISTRIBUTION Often planted in other habitats for withies, masking its true native range.

Eared Willow
■ *Salix aurita* Height >2m

DESCRIPTION Shrubby, much-branched and small willow. Shoots are downy at first, and become shiny and brown with age. Broadly ovate leaves are >4cm long, with wavy margins and a twisted tip, reddish petioles, and large, leafy stipules (ears) at the leaf base. Male catkins are ovoid and yellow. Female catkins are greener, borne on separate trees and appear just before the leaves. **HABITAT** Native species favouring damp, acid soils. Common beside moorland and upland streams, and on damp heaths. **DISTRIBUTION** Widespread, sometimes appearing in areas newly cleared by forestry operations.

Grey Willow ■ *Salix cinerea* Height >6m

DESCRIPTION Variable; usually a large shrub or sometimes a small tree with

characteristic thick, downy grey twigs. Represented by ssp. *cinerea* and ssp. *oleifolia* (syn. *S. c. atrocinerea* or *S. atrocinerea*). If bark is peeled from two-year-old twigs, wood shows a series of fine, longitudinal ridges. Leaves are oblong and pointed, usually about four times as long as broad, on short petioles with irregular stipules. They often have inrolled margins and are grey and downy below. Upper surface is matt and downy in ssp. *cinerea*, but glossy and hairless in ssp. *oleifolia*. Catkins appear in early spring on separate trees, before leaves. Male catkins are ovoid and yellow. Female catkins are similar but greener, and eventually release finely plumed seeds. **HABITAT** Native species usually found in wet habitats such as fenlands, stream-sides and damp woodland. **DISTRIBUTION** Common across much of N Europe.

Woolly Willow
■ *Salix lanata* Height >3m

DESCRIPTION Very distinctive
small shrub with pale, woolly
young shoots, which become glossy
brown with age. Leaves are broadly
oval, >6cm long, white and woolly
below when mature. Male and
female catkins are produced on
separate tree. Males catkins are
golden, female paler and very hairy.
HABITAT Mountain ledges and
stream-sides, especially on base-rich
rocks. DISTRIBUTION Restricted
in the wild to mountainous
habitats in the far north, but a
popular subject for alpine gardens
elsewhere.

Almond Willow
■ *Salix triandra* Height >10m

DESCRIPTION Small tree, although more
often an untidy shrub with smooth bark
that flakes off in small patches. Greenish or
reddish-brown shoots tend to snap easily,
and terminate in brown, ovoid, smooth
buds. Leaves are ovate, >10cm long, each
with a serrated margin and pointed tip,
usually smooth, hairless and dark glossy
green. Petiole is >1.5cm long and smooth.
Catkins appear at the same time as leaves
on short, leafy shoots, and are erect and
cylindrical. Male catkins, >5cm long and
greenish-yellow, remain on tree for some
time. Female catkins are shorter and
more compact. HABITAT Native species
growing in damp ground, often beside
rivers and ponds, or in marshes. Planted
elsewhere for biofuel and basket weaving.
DISTRIBUTION Widespread in S England
and southern mainland Europe, but
becomes scattered further north and west.

Goat Willow or Sallow ■ *Salix caprea* Height >12m

DESCRIPTION May grow as a multi-branched, dense, shrubby tree, or a taller tree with a straight, ridged stem and sparsely domed crown. Thick, stiff twigs are hairy at first becoming smoother and yellowish-brown with age. Large, oval leaves, >12cm long, each

have a short, twisted point at the tip. Upper leaf surface is dull green; lower surface is light grey and woolly. Male and female catkins grow on separate trees, and appear before leaves. Measuring >2.5cm long, they are ovoid and covered with greyish, silky hairs before opening. At this time the tree is often called Pussy Willow because the silky-grey buds bear a fanciful resemblance to cats' paws. Open male catkins become bright yellow. Female catkins are greener and produce numerous silky-haired seeds. HABITAT Native species growing in woods, hedgerows and scrub, often in drier places than other similar species. DISTRIBUTION Common and widespread, and an important tree for wildlife.

Hoary or Olive Willow
■ *Salix elaeagnos* Height >6m

DESCRIPTION Similar in form to Common Osier (see p. 59). Young twigs have dense grey or white hairs; older twigs become yellow-brown and smooth. Best recognized by studying the leaves, which have matt white hairs beneath and are dark shiny green above when mature. Leaves are >15cm long and <1cm wide, and have untoothed margins. Male and female catkins appear on separate trees just before leaves; reddish male catkins are >3cm long, female catkins are smaller. HABITAT Native to continental Europe, favouring open, sunny sites, mountain slopes, riversides and ditches, and usually avoiding shade. DISTRIBUTION Sometimes planted for ornament in Britain, where it has become naturalized in places.

Cricket-bat Willow
■ *Salix alba* var. Caerulea Height >25m

DESCRIPTION Large, broadly columnar tree with dark grey bark and a dense crown when growing in the open, but more slender and upright when crowded. Yellowish-grey shoots are downy at first. Bluish-grey leaves are almost hairless. Trunk is straight with purplish-red shoots. Male catkins are small, elongated-ovoid and pendulous. Female catkins are longer, slender, green and borne on separate trees. **HABITAT** Damp lowlands, riversides, marshes and fens. **DISTRIBUTION** Easily raised from cuttings and grows well in damp soil, especially near ponds and rivers. Planted as source of timber for cricket bats.

Western Balsam-poplar ■ *Populus trichocarpa* Height >35m

DESCRIPTION Fast-growing (>2m per year) tree, columnar when mature, with a tapering crown and trunk. Bark is dark grey with shallow grooves and fissures. Leaves are pointed, tapering, glossy green above and white below, and turn yellow in autumn. Catkins are slender and pendulous; male catkins are reddish-brown, female greenish. Seeds are hairy and produced abundantly. **HABITAT** Native to Pacific coast of North America. Favours wetland margins, mountain slopes and valleys, and riversides. **DISTRIBUTION** Planted in Europe, thriving in wetter areas. Sometimes planted for pulpwood or plywood production.

White Poplar or Abele
■ *Populus alba* Height >20m

DISTRIBUTION Easily identified tree in windy weather in summer, when the pure white undersides of the leaves are turned up and the whole tree looks white. White bark on trunk in mature specimens is broken by diamond-shaped scars. Shoots are covered in white felt, but this usually wears off by the end of the growing season. Simple, deeply lobed leaves are covered with dense white felt underneath, but are greyish-green above. Male catkins are long, ovoid, white and fluffy, female more slender and greenish-yellow. **HABITAT** Native to mainland Europe, favouring wet areas, coasts, riversides, fens and marshes, and often growing in thickets due to suckering. **DISTRIBUTION** Presumed to have been introduced to Britain (one of the first tree introductions, in the 16th century), perhaps because of its rapid rate of growth and ability to flourish even in the poorest of soils and on the most exposed sites.

Grey Poplar
■ *Populus x canescens* Height >37m

DESCRIPTION Hybrid between White Poplar and Aspen (see above and p. 65), which grows into a very large tree with a solid bole. Whitish appearance when the wind turns the leaves in spring, but not as strikingly white as White Poplar. Whitish bark has diamond-shaped fissures. Rounded to oval leaves, borne on long petioles, are toothed with regular blunt, forwards-pointing teeth. Leaf upper surface is glossy grey-green; lower surface is covered with a greyish-white felt that is lost by midsummer. Male and female catkins are borne on separate trees. Female trees with green, pendulous catkins are rare. Male catkins are elongated and pendulous, giving the whole tree a purplish colour when they swell before opening in spring. **HABITAT** Native to mainland Europe. Tolerant of a wide range of soils and climates. **DISTRIBUTION** Introduced into Britain very early, probably with White Poplar.

Aspen

■ *Populus tremulus* Height >18m

DESCRIPTION Slender to slightly conical tree with a rounded crown and tall, tapering trunk. Best known for its fluttering leaves, which rustle in the slightest breeze. Smooth, greyish-green bark becomes brown, ridged and fissured with age. Leaves are rounded to slightly oval, with shallow marginal teeth, green on both surfaces, but paler below, on long, flattened petioles. Leaves newly produced in summer are often deep red. Catkins, >8cm long, are produced in clusters at ends of twigs, with males and females on different trees. Male catkins are reddish-purple, female green tinged pink. **HABITAT** Native to Europe, growing in woodland, marshes, riversides and hedgerows, often on very poor soils. **DISTRIBUTION** Common in many places, especially on poor, damp soils, and able to grow well north of Arctic Circle.

Black Poplar

■ *Populus nigra* ssp. *betulifolia*
Height >32m

DESCRIPTION Large, spreading tree when fully mature, with a domed crown and thick, blackish, gnarled bole covered with distinctive burrs and tuberous growths. Shoots and buds are smooth and golden-brown when young. Triangular, long-stalked leaves have a finely toothed margin and are shiny green on both surfaces. Male catkins are pendulous and reddish, female catkins greenish. **HABITAT** Native to Europe, thriving best on heavier soils, and favouring riverbanks, marshes, fens and damp woodland. **DISTRIBUTION** Widespread in Europe, and sometimes planted in urban parks because of its tolerance to pollution.

Lombardy Poplar

■ *Populus nigra* 'Italica' Height >36m

DESCRIPTION Distinctive, narrowly columnar and very tall tree. Gnarled bole supports numerous short, ascending branches that taper towards the narrow, pointed crown. Otherwise similar to Black Poplar (see p. 65), but with slightly more triangular leaves. Typical, slender Lombardy Poplars are all males, bearing reddish catkins. Female tree, known as var. 'Gigantea', is scarce and has thicker, spreading branches that give the tree a broader crown. **HABITAT** Native to Italy, and tolerates a wide variety of soils and climates, but usually occurs in lowland areas. **DISTRIBUTION** Introduced to Britain and elsewhere in Europe in the mid-18th century. Often planted in long lines.

Hybrid Black Poplar
■ *Populus* x *canadensis* Height >30m

DESCRIPTION Upright or spreading tree, depending on situation, with a narrow crown. Similar to Black Poplar (see p. 65), one of its parent species, and in many areas far more common; the other parent is the North American Cottonwood. Trunk lacks burrs seen in Black Poplar, but has deeply fissured and greyish bark. Young twigs are greenish or slightly reddened. Leaves are alternate, oval to triangular, and sharply toothed with fringes of small hairs. Catkins are similar to those of Black Poplar. Many hybrid forms occur, separated by leaf structure and tree shape, for example Black Italian Poplar. **HABITAT** Lowland areas, avoiding very wet soils; not common in colder northern climates. **DISTRIBUTION** Widely planted for ornament and timber (used for packing crates and boxes).

Balm-of-Gilead ■ *Populus x jackii (candicans)* Height >20m

DESCRIPTION A notable feature of this tree is the abundant sticky buds, on downy shoots, that are balsam scented. Young leaves are also aromatic when newly opened, and heart shaped and downy below, on a downy petiole. Mature tree is open crowned, freely sending up suckers that form thickets. Male catkins are pendulous and yellowish-green. Female catkins produce an abundance of cottony white seeds. **HABITAT** Naturally occurring hybrid between Eastern Balsam Poplar (see below) and **Eastern Cottonwood** *P. deltoides*, found mainly in E USA. Tolerates a wide range of soils and climate, but thrives in wet, well-drained and fairly open sites. **DISTRIBUTION** Often planted in Britain and mainland Europe as an ornamental tree.

Eastern Balsam Poplar ■ *Populus balsamifera* Height >30m

DESCRIPTION Conical to slightly spreading tree with numerous ascending branches arising from a tapering bole; base of bole is often surrounded by suckers. Bark is thinner than that of other poplars, and narrowly grooved. Young shoots and 2.5cm-long buds are covered with shiny resin. Leaves are >10cm long, oval and pointed at the tips with finely toothed margins; they are dark shiny green above, paler and downy below. Greenish catkins appear in late spring or early summer; males >7.5cm long and females >12.5cm long grow on separate trees. **HABITAT** Native to North America, growing in river valleys, and on floodplains and mountain slopes. **DISTRIBUTION** Occasionally cultivated in Europe. All cultivated trees appear to be males.

Shagbark Hickory

■ *Carya ovata* Height >20m

DESCRIPTION Upright or slightly spreading tree with a broad, flattened crown. Grey bark splits into long, scaly flakes. In winter sparse branches support reddish twigs tipped with scaly buds. Compound leaves have five, or sometimes three or seven, leaflets >20cm long, longer near tip of leaf. Each leaflet is oval to oblong and toothed, with tufts of short white hairs between teeth. Terminal leaflet is short stalked. Leaves are leathery and sometimes oily. Male catkins are green, >15cm long and grow in spreading clusters. Female flowers are small, yellowish and grow in terminal clusters. Fruits are round, >6cm long and occur on short stalks; they contain white seeds. **HABITAT** Native to E North America, growing in forest margins, woodland and hillsides. **DISTRIBUTION** Occasionally planted in our area.

Bitternut ■ *Carya cordiformis* Height >30m

DESCRIPTION Large deciduous tree with a high, conical crown. Bark is greyish, smooth at first and becoming scaly with age; peeling flakes reveal orange patches beneath. Branches are mostly straight and ascending, the greenish twigs tipped with elongated,

yellowish, scaly buds. Leaves are compound, with nine leaflets. Individual leaflets are elongated, pointed at the tip and have toothed margins. Male catkins are >7cm long, yellowish and pendulous. Fruits, >3.5cm long, are rounded to pear shaped with four wings, concealing grey, smooth seeds. **HABITAT** Native to E North America, growing in humid mountain valleys, swamps and riverbanks. **DISTRIBUTION** Planted in our area mainly for ornament.

Caucasian Wingnut
■ *Pterocarya fraxinifolia* Height >35m

DESCRIPTION Spreading tree with a domed crown and stout bole, from which many branches arise close to the same point and suckers grow freely. Bark is grey, fissured and gnarled. Compound leaves have 11–20 pairs of leaflets, each >18cm long, ovate to lanceolate, and with a pointed tip and toothed margins. Midribs bear stellate hairs on undersides. Leaves turn yellow in autumn. Male catkins are solitary, female pendent with many flowers giving rise to broad-winged nutlets.
HABITAT Native to SW Asia. Favours riverbanks, floodplains and marshes with deep soil and mild climates.
DISTRIBUTION Planted outside its native range for ornament.

Common Walnut ■ *Juglans regia* Height >30m

DESCRIPTION Spreading deciduous tree with domed crown and straight bole when grown in ideal conditions; often contorted when found in orchards. Bark is smooth and brown at first, becoming grey and fissured with age. Lowest branches are spreading to ascending, and often large near base, but dividing rapidly into numerous twisted twigs with dark purple-brown buds. Compound leaves with 7–9 elliptical leaflets, >15cm long,

thick and leathery, with pointed tips and untoothed margins. Crushed leaves are slightly aromatic. Male catkins are yellow and >15cm long; female flowers are small and greenish, with yellow protruding, branched stigma. Fruits are rounded, >5cm long; smooth green skin, dotted with slightly raised glands, encases the familiar edible walnut seed.
HABITAT Native to SE Europe and Asia, growing in woods, on hillsides and in lowland grasslands. **DISTRIBUTION** Long cultivated in other parts of Europe, and may have arrived with the Romans. Rarely naturalized.

Black Walnut ■ *Juglans nigra* Height >32m

DESCRIPTION Tree with a tall, straight bole and domed crown of brighter green leaves than those of Common Walnut (see p. 69). Dark brown bark splits to show a diamond pattern of deep cracks. Leaves are compound with 15–23 leaflets, finely toothed and downy below. Fruits are similar to Common Walnut's, but not as edible; green husk yields a similar dark dye. **HABITAT** Native to central and E USA. Grows in woodland, meadows, riversides and open, sunny sites in lowland areas. **DISTRIBUTION** Occasionally planted in Europe as an ornamental tree. Yields valuable oils and timber for furniture.

Butternut

■ *Juglans cinerea* Height >26m

DESCRIPTION Tall, slender tree with rather smooth, light grey bark. Compound leaves are >70cm long, and the leaflets are more widely spaced than in Black Walnut (see above). Central leaf stalk is densely hairy and the leaflets near the leaf base are the smallest. Lemon-shaped, edible fruits form in clusters of 3–6. A fast-growing but rather short-lived tree (75 years), susceptible to some diseases. **HABITAT** Native to E North America, favouring riversides, meadows, rocky hillsides and woodland margins. **DISTRIBUTION** Planted in Europe in parks and gardens for ornament.

Bog Myrtle

■ *Myrica gale* Height >1– 2m

DESCRIPTION Woody, brown-stemmed deciduous shrub with 2–5cm-long, oval, grey-green and resin-scented leaves that grow spirally on the twigs. Orange, ovoid male catkins and pendulous brown female catkins appear in spring on separate plants. Fruits are brownish nuts. Roots contain nodules of nitrogen-fixing bacteria. Scented leaves are thought to have medicinal properties and have been used as insect repellents. **HABITAT** Boggy heathland, edges of moorland pools and streams, tolerating very acidic conditions. **DISTRIBUTION** Widespread but local species in suitable habitats in N Europe.

Dwarf Birch

■ *Betula nana* Height >1m

DESCRIPTION Low-growing and often rather prostrate undershrub with short and upright or spreading branches, and stiff, hairy twigs and reddish-brown bark. Leaves are rather rounded, 6–8mm across, and coarsely toothed and hairy when young, smooth and hairless when mature. Female catkins sit upright on the twigs and are covered with three-lobed scales. Male catkins are yellow and pendent. **HABITAT** Northern, tundra species found in uplands in south of its range, but near sea level in Arctic regions. Occurs on upland heaths and mountain slopes, and in bogs. Tolerant of both waterlogged and relatively free-draining, peaty soils. **DISTRIBUTION** Sometimes grown in rock gardens because of its small size and compact habit.

Silver Birch ▪ *Betula pendula* Height >26m

DESCRIPTION Slender deciduous tree with a narrow, tapering crown. Older trees acquire a weeping habit and have thick bark, deeply fissured at base of bole, and breaking up into rectangular plates; higher up the bark is smooth silvery-white, often flaking away. Branches are ascending in young trees, but twigs and shoots are pendulous, smooth, brownish and pitted with white resin glands. Leaves are >7cm long, triangular, pointed, toothed, thin and smooth when mature; they grow on hairless petioles and turn golden-yellow in autumn. Male catkins are produced in groups of 2–4 at tips of young twigs appearing early in winter; when leaves are opening, they are yellow and pendulous. Female catkins are shorter, more erect, greenish and produced in axils of leaves. Seeds are winged and papery and usually produced copiously. **HABITAT** Native to wide area of Europe, including Britain and Ireland. Wide range of habitats and soil types, but thrives on open sites, hillsides, heathland and woodland margins. **DISTRIBUTION** Often planted as an ornamental tree in gardens and parks; readily colonizes new areas.

Downy Birch
▪ *Betula pubescens* Height >25m

DESCRIPTION Superficially similar to Silver Birch (see above), with which it often grows, but the crown in winter looks untidy by comparison. A rather variable species, but usually easy to recognize by soft, downy feel of tips of twigs in spring, and reddish bark on young wood. Bark does not break up into rectangular plates at the base like that of Silver Birch. Branches are also more irregular and densely crowded, and mostly erect, never pendulous. Twigs lack the whitish resin glands, but do have a covering of downy white hairs. Catkins are similar to Silver Birch's, but seeds have smaller wings. **HABITAT** Native to most of Europe, including Britain and Ireland. Common in upland areas and on poor soils, and rapidly colonizes cleared woodland and heaths. **DISTRIBUTION** Sometimes planted as a fast-growing timber tree.

Himalayan Birch
▪ *Betula utilis* Height >20m

DESCRIPTION Attractive deciduous tree with extremely colourful bark, gleaming white in var. *jacquemontii*, but pink, red or golden in other forms. Bark is marked with horizontal lenticels and peels off horizontally into rolls. Branches are mainly upright rather than spreading, with twigs that are hairy when young. Leaves are oval, each with a pointed tip and toothed margins, and dark green with 7–14 pairs of veins depending on the variety. Male catkins are long, pendulous and yellow; female are smaller and green. **HABITAT** Native to the Himalayas, growing on high mountain slopes and in forests. **DISTRIBUTION** Planted in our area as an ornamental, mainly for its stunning bark, which is particularly striking and evident on a sunny winter's day.

Erman's Birch
▪ *Betula ermanii* Height >24m

DESCRIPTION Fast-growing and attractive tree easily recognized by studying its bark, which is pinkish, or sometimes shining yellowish-white. It peels horizontally and hangs in tattered strips down the boles of mature trees; younger trees have smoother white bark. More spreading and with a stouter bole than Silver Birch (see p. 72), with which it frequently hybridizes. Mostly upright branches have twigs that are warty and usually hairless. Triangular to heart-shaped leaves each have a pointed tip and toothed margins with 7–11 pairs of veins and hairless stalks. **HABITAT** Native to E Asia, growing on hillsides, and in river valleys and woodland margins. **DISTRIBUTION** Introduced into Europe from Japan. Best specimens seen in established gardens, where peeling bark is a fine winter feature.

Common Alder

■ *Alnus glutinosa* Height >25m

DESCRIPTION Small, spreading and sometimes multi-stemmed tree with a broad, domed or conical crown, and brownish bark fissured into square or oblong plates. Branches are ascending in young trees, and spreading later. Twigs are smooth except when young, when they have a sticky feel (hence *glutinosa*), and raised orange lenticels. Leaves are stalked and rounded, >10cm long, and with a slightly notched apex, and wavy or bluntly toothed margin. The 5–8 pairs of veins have long hairs in axils on the underside of the leaf. In winter purplish male catkins, in bunches of 2–3, are an attractive feature; in spring they open, revealing yellow anthers. Female catkins are smaller (1.5cm) and cone-like, reddish-purple at first, then turning green, usually in bunches of 3–8. They form hard green 'cones' that persist until the following spring. Their small, winged seeds float on water, aiding dispersal. **HABITAT** Native to Europe, including Britain and Ireland; absent only from very far north. Favours riversides, marshes and wet woodland on heavy soils. **DISTRIBUTION** Occurs throughout Europe.

Green Alder ■ *Alnus viridis* Height >5m

DESCRIPTION Rarely more than a large shrub or small tree, with brown bark and mostly smooth, greenish twigs, and pointed, sessile, shiny red buds. Leaves are more pointed than those of Common Alder (see above), sharply toothed, and hairy on the midrib and in junctions of veins on undersides. When first open they are sticky to the touch. Male

catkins are >12cm long, yellow and pendulous; female catkins 1cm long, erect and greenish at first, becoming reddish later, and usually found in stalked clusters of 3–5. Cone-like ripe catkins are rounded, green and tough at first, becoming blackened later; they persist until the following spring. **HABITAT** Native to mountains of central and E Europe, favouring mountain slopes, riversides and woodland margins. **DISTRIBUTION** Planted outside native range for ornament.

Grey Alder ■ *Alnus incana* Height >25m

DESCRIPTION Fast-growing alder more at home on dry soils than most other alders. Shoots and new leaves are covered with a dense layer of soft greyish hairs (hence *incana*). Bark is smooth and grey. Leaves are triangular and toothed, terminating in a point, and margins do not roll inwards. Hairs persist on underside of leaf as it matures. Catkins and fruits are very similar to those of other alders, although the green fruits are more globose before ripening to the typical dark, woody alder cone. **HABITAT** Native to mainland Europe, on hillsides, stony ground and cliffs, and in rocky areas. **DISTRIBUTION** Introduced into Britain but not often planted. Good species for wasteland and reclamation schemes.

Italian Alder
■ *Alnus cordata* Height >29m

DESCRIPTION Attractive tree with a bold, conical shape, fine, glossy leaves and an impressive show of catkins and cones. Pale grey bark is fairly smooth with slightly downy twigs. Key feature for identification is the glossy, heart-shaped leaves (hence *cordata*), which have short tufts of orange hairs along midrib on underside. Male catkins are yellow and produced prolifically; female catkins are borne in small clusters, ripening in early summer. Woody 'cones' are larger than those of any other alder species. **HABITAT** Native to Corsica and S Italy, on hillsides and rocky slopes, and in woodland margins. **DISTRIBUTION** Planted outside native range in parks and gardens, and often along roadsides.

Hornbeam ■ *Carpinus betulus* Height >30m

DESCRIPTION Fine tree with a striking outline in winter. Bole is often gnarled and twisted, and bark is silvery-grey with deep fissures lower down and occasional dark bands. Branches are usually densely packed, ascending and twisted, bearing greyish-brown, partly hairy twigs. Leaves are oval and pointed with a rounded base, short petiole and double-toothed margin; the 15 pairs of veins are hairy on the underside. Trees planted in hedgerows retain their leaves long into winter. Male catkins, >5cm long, are yellowish-green with red outer scales. Fruits are produced in clusters of winged nutlets, >14cm long, usually consisting of about eight pairs of small, hard-cased nuts with a three-pointed, papery wing. **HABITAT** Native to Britain and mainland Europe, growing in woodland and hedgerows on a variety of soil types. **DISTRIBUTION** Widely planted and seen as a specimen tree in parks and gardens, and often planted for its tough timber.

European Hop-hornbeam
■ *Ostrya carpinifolia* Height >19m

DESCRIPTION Spreading tree with a domed crown and robust bole. Grey-brown bark breaks up into squarish plates. Branches are almost level when growing in the open, but in a woodland they may be more crowded and ascending. Leaves are similar to Hornbeam's (see above), being oval with a toothed margin. Yellowish-green catkins appear in spring. Fruits form in clusters and have a superficial resemblance to bunches of hops; they may persist through winter. **HABITAT** Native to mainland Europe, on open hillsides and in mixed deciduous woodland. **DISTRIBUTION** In Britain occurs mainly in well-established gardens and tree collections.

Hazel ■ *Corylus avellana* Height >6m

DESCRIPTION Often no more than a spreading, multi-stemmed shrub, but sometimes grows into a taller tree with a shrubby crown and short but thick and gnarled bole. Bark is smooth and often shiny. Branches are upright to spreading, and twigs are covered with stiff hairs; the oval buds are smooth. Leaves are rounded, >10cm long, each with a heart-shaped base and pointed tip. Margins are double-toothed and upper surface is hairy. Leaf veins have white hairs on undersides. Petiole is short and hairy, and the whole leaf has a bristly, rough feel. Male catkins first appear in autumn and are short and green, but in spring they are >8cm long, pendulous and yellow. Female flowers are red and very small, and produce hard-shelled nuts in bunches of 1–4; nuts are partly concealed in a leafy, deeply toothed involucre. Nut is up to 2cm long, brown and woody when ripe. **HABITAT** Native to Europe, including Britain and Ireland, in woodland, hedgerows and scrub, and on hillsides. **DISTRIBUTION** Widespread across most of Europe, and often planted for its useful timber.

Filbert
■ *Corylus maxima* Height >6m

DESCRIPTION Very similar tree to Hazel (see above), but with slightly larger leaves and larger nuts, which are mostly solitary or in bunches of 2–3, and entirely enclosed in an undivided involucre, which is constricted over the nut and toothed at the tip. A cultivated form with purple leaves is sometimes seen in parks and gardens. **HABITAT** Native to Balkans, growing on hillsides, and in stony areas, woodland and dry valleys. **DISTRIBUTION** Widely planted outside its native range, for example in SE England, for superior quality of its nuts; sometimes naturalized.

Common Beech ▪ *Fagus sylvatica* Height >40m

DESCRIPTION Large and imposing deciduous tree with a broad, rounded crown and smooth and grey bark. Buds are >2cm long, smooth, pointed and reddish-brown. Leaves

are 10cm long, oval and pointed, with a wavy margin and a fringe of silky hairs when freshly open. Male flowers are pendent and borne in clusters at tips of twigs. Female flowers are paired, borne on short stalks and surrounded by a brownish, four-lobed involucre. Nuts are >1.8cm long, three sided, shiny and brown, and enclosed in a prickly case in pairs. **HABITAT** Native to central and W Europe, including Britain, occurring in woodland and hedgerows, and on hillsides. Prefers drier alkaline soils, but can tolerate more acid conditions. **DISTRIBUTION** Widespread and widely planted as an ornamental and timber-producing tree.

Oriental or Eastern Beech ▪ *Fagus orientalis* Height >23m (rarely 30m)

DESCRIPTION Very similar tree to Common Beech (see above), with smooth bark and a stout bole, but leaves are larger and widely separated, with seven or more pairs of veins. Small catkins appear in spring just after leaves, and in autumn small, triangular nuts form in pairs in smooth, leafy capsules. May hybridize with Common Beech. Grows vigorously and forms a fine tree in good conditions. **HABITAT** Native to Balkans, Asia Minor and the Caucasus. Grows on hillsides and mountain slopes, and in forests, usually on lime-rich soils, but can tolerate other conditions. **DISTRIBUTION** Rare outside its native range.

Rauli or Southern Beech ■ *Nothofagus procera* Height >28m

DESCRIPTION Attractive conical tree with a stout bole and striking autumn foliage. Grey bark splits into vertical plates. Lower branches are usually level, upper branches more ascending. Thick green twigs that darken with age bear pointed, reddish brown buds.

Leaves are alternate, >8cm long and pointed with wavy margin. Male and female flowers occur on the same tree. Female flowers are solitary and give rise to four-lobed, hairy capsules containing three shiny brown nuts. Grows rapidly at first, >2m a year, and soon makes an attractive specimen tree. **HABITAT** Native to Chile. Grows on mountain slopes and hillsides in a variety of soil types. **DISTRIBUTION** Introduced to our area in the 20th century, and found in parks, gardens and commercial plantations.

Roble Beech

■ *Nothofagus obliqua* Height >30m

DESCRIPTION More delicate in appearance than Rauli (see above), with light grey, peeling bark, and slender, ascending branches with pendent shoots on upper crown. Twigs are finer than those of Rauli, and branch in a regular, alternate pattern. Leaves have 7–11 pairs of veins (compared with Rauli's 15–22), and wavy margins. Flowers grow in leaf axils, and fruits are four-lobed, hairy capsules. **HABITAT** Native to Chile and W Argentina. Grows on mountain slopes and hillsides, and on stony ground, but tolerant of a range of soils and climates. **DISTRIBUTION** Grown in our area for ornament, having fine autumn colour, and occasionally planted for timber.

Antarctic Beech
■ *Nothofagus antarctica* Height >16m

DESCRIPTION Forms an attractive small tree with delicate shiny foliage and reddish, shiny bark in young trees. Leaves have only four pairs of veins and remain curled for most of the season, turning a pleasing yellow, then brown in autumn. Young leaves have a pleasant scent. Tiny yellow catkins emerge in spring, and lightly scented nuts form in a capsule in autumn. **HABITAT** First discovered in its native Chile and Tierra del Fuego in the 1830s. Grows on mountain slopes and hillsides in areas of high rainfall. The most southerly tree species in the world. **DISTRIBUTION** Grown in Britain since its discovery. Hardy, but prefers some shelter.

Sweet Chestnut
■ *Castanea sativa* Height >35m

DESCRIPTION Handsome, large-leaved deciduous tree with a fine bole and attractive autumn colours. Young trees have smooth grey bark; spiral fissures in mature trees. Glossy leaves are >25cm long and lanceolate, with margins serrated with spine-tipped teeth; leaf is pointed at the tip and sometimes has a slightly heart-shaped base. Male catkins are creamy-white, long and pendulous, producing a sickly sweet smell. Female flowers are borne in groups of two and three at bases of male catkins, giving rise to familiar prickly green capsule that splits to release three shiny, brown-skinned nuts. **HABITAT** Native to most of mainland Europe (not Britain and Ireland) and North Africa. Grows on hillsides and mountain slopes, and in woodland, preferring well-drained, slightly acidic soils. **DISTRIBUTION** Widespread in Britain following introduction by Romans; now frequently planted and sometimes naturalized.

English or Pedunculate Oak
■ *Quercus robur* Height >36m

DESCRIPTION Large, spreading deciduous tree with a dense crown of heavy branches, and thick grey, fissured bark. Very old trees (700–800 years) may have dead branches emerging from upper canopy and a hollow trunk. Shoots and buds are hairless. Deeply lobed leaves have two auricles at bases; borne on very short stalks (5mm or less). First flush of leaves is often eaten rapidly by insects, and replaced by a second crop in midsummer – so-called 'Lammas growth'. Male and female catkins are produced in spring. Acorns are borne on long stalks in roughly scaled cups, in groups of 1–3. **HABITAT** Widespread native tree in Britain and Ireland. Lowland areas, hillsides, woodland and hedgerows, especially on heavier clay soils. **DISTRIBUTION** Often dominant in old woodland, especially in lowland areas, but occurs in more hilly country as well. Frequently planted and of immense value for wildlife and as a timber tree.

Sessile Oak ■ *Quercus petraea* Height >40m (rarely 43m)

DESCRIPTION Sturdy deciduous tree with a domed shape and relatively straight branches radiating around a longer and more upright bole than that of English Oak (see above).

Grey-brown bark has deep vertical fissures. Buds are orange-brown with long white hairs. Lobed, flattened leaves are dark green with hairs along veins; they are produced on yellow stalks, 1–2.5cm long, and lacking auricles at bases. Male catkins appear in spring. Acorns are long, egg shaped and stalkless. **HABITAT** Hilly and rocky areas on poor soils; tolerant of high rainfall. **DISTRIBUTION** Most common in western areas of Britain and Ireland, and previously coppiced for bark (tanning) and fuel wood. Supports many insect species.

Downy Oak ■ *Quercus pubescens* Height >24m

DESCRIPTION Similar to English Oak (see p. 81), forming a large, sturdy tree under good growing conditions. Deep grey bark is grooved with numerous deep fissures and small plates or rough scales. Twigs and buds are covered with greyish downy hairs, and

buds look orange-brown beneath the down. Leaves are smaller than English Oak's, >13cm long and 6cm wide, with shallower, forwards-pointing lobes and very hairy petioles. Young leaves are densely downy at first, but become smoother and grey-green above when mature. Catkins appear in late spring. Acorns are sessile, borne in stalkless, shallow cups about 1.5cm deep and covered in closely packed, downy scales. **HABITAT** Native to Europe. Grows on hillsides and in woodland on dry, lime-rich soils, generally avoiding very wet or cold areas. **DISTRIBUTION** Occasionally planted outside its native range as an ornamental tree.

Hungarian Oak
■ *Quercus frainetto* Height >30m

DESCRIPTION Deciduous rapidly growing oak that forms a fine, broadly domed tree with pale grey, finely fissured bark, breaking into fine ridges. Largest branches are long and straight, emerging from a sturdy bole, and terminating in finely downy, greyish-green or brownish twigs. Large, deeply lobed leaves are >25cm long and 14cm wide. Pendulous yellow catkins appear in late spring; acorns are borne in cups about 1.2cm deep covered in downy, blunt, overlapping scales. **HABITAT** Native to Balkans, central Europe and S Italy. Grows in drier areas on acid soils, avoiding riversides and wet valleys. **DISTRIBUTION** Planted outside its native range for its splendid appearance when mature. Sometimes grafted onto stock of English Oak.

Cork Oak ■ *Quercus suber* Height >17m

DESCRIPTION Medium-sized evergreen oak forming a rounded tree. Striking bark is thick, and pale greyish-brown with deep fissures and ridges if left to mature, with a soft, corky texture. Numerous large, twisted branches arise low down on bole; in very old trees some may trail on the ground. Leaves resemble holly leaves, with spiny tips to shallow lobes. Mature leaves are dark green and smooth above, but paler, almost grey and downy below. Acorns are 2–3cm long, egg shaped and borne in cups covered with long, projecting scales. **HABITAT** Native to Mediterranean region. Grows in valleys, and on hillsides and plains, mainly in lowland areas, often forming extensive woodland. **DISTRIBUTION** Introduced outside its native range and grown for ornament as far north as Scotland. Bark is harvested for corks for wine bottles.

Turkey Oak ■ *Quercus cerris* Height >38m

DESCRIPTION Deciduous broadly conical oak, becoming more spreading and domed with age. Thick, grey-brown bark becomes fissured, forming regular, square-shaped plates in older trees. Branches appear swollen near base and spread upwards. Buds are covered with long hairs. Leaves are deeply lobed with >10 lobes or large teeth, on 1–2cm-long, slightly downy petioles. Upper leaf surface is deep green; lower surface is downy when new, and greyish. Catkins appear in late spring. Acorns are partly encased in a deep cup covered in long, outwards-pointing scales. **HABITAT** Native to S Europe, growing on hillsides and in valleys on a wide range of soils, and often forming extensive woodland. **DISTRIBUTION** Introduced into Britain by J. Lucombe of Exeter in 1735; now widely planted in parks and gardens and sometimes occurs in woodland.

Lucombe Oak ■ *Quercus x hispanica* 'Lucombeana' Height >35m

DESCRIPTION Tall evergreen hybrid between Cork and Turkey Oaks (see p. 83). Bark is variable; some specimens are similar in character to Cork Oak, while others have

smoother, darker bark. Leaves are long, glossy and toothed, and remain on tree throughout all but the hardest winters. Some of the earliest trees, dating from the original hybridization, lose a large proportion of their leaves; later crosses have a more dense crown. Male catkins are produced in early summer, and acorns form in small, scaly cups. **HABITAT** Most trees are seen in parks and gardens in milder areas of Britain, especially near the sea. **DISTRIBUTION** Hybrid originated in Exeter, Devon, in the 18th century, and was named after Lucombe's nursery. Still most common in parks and gardens around Exeter, but may also be found in mature parks and gardens in sheltered regions elsewhere.

Holm or Evergreen Oak ■ *Quercus ilex* Height >28m

DESCRIPTION Broadly domed tree often with a dense, twiggy crown. Very dark bark has shallow fissures, eventually cracking to form squarish scales. Branches appear from

low down on bole; young shoots are covered with white down. Leaves are variable, but usually ovate to oblong with a pointed tip and rounded base on mature trees, but more like holly leaves on a young tree, dark green and glossy. Male catkins appear in spring, their golden colour contrasting with silvery new leaves. Acorns sit deeply in cups covered with rows of small, hairy scales. **HABITAT** Native to S Europe, growing in drier areas on well-drained soils, often near coast, mostly in S Europe. **DISTRIBUTION** Widely planted outside its native region, especially as a shelter belt in coastal regions.

Red Oak

■ *Quercus rubra* Height >35m

DESCRIPTION Broadly conical tree with pale silvery-grey, sometimes brownish bark, which is mostly smooth but fissured with age. Leaves are large, usually 10–20cm long and deeply lobed, with smaller teeth terminating in fine hairs at tips of lobes. They are green above and paler matt green below during the growing season, and turn red or brown in autumn. Young trees produce the finest red colouring. Pendulous male catkins appear in spring, turning the tree golden-yellow. Acorns are rounded, and grow in neat, scaly cups. **HABITAT** Native to North America. Grows in lowland areas and on hillsides with rich, mildly acid soils. **DISTRIBUTION** Planted in Europe for its spectacular autumn colours; occasionally naturalized.

Pin Oak ■ *Quercus palustris* Height >26m

DESCRIPTION Broadly conical deciduous tree with a short bole and smooth, grey-brown bark. Branches are mostly ascending. Distinctive leaves are >12cm long and deeply lobed, with bristles at tips of pointed lobes. They are glossy green on both surfaces in summer, and palest below; there are tufts of brownish hairs in the vein axils. Male catkins are pendulous, yellowish and open in early summer. Acorns >1.5cm long are partially enclosed in shallow, scaly cups. **HABITAT** Native to E North America. Grows in riversides, valleys, lake margins and swamps, avoiding lime-rich soils. **DISTRIBUTION** Introduced to Europe as an ornamental tree.

Scarlet Oak
■ *Quercus coccinea* Height >28m

DESCRIPTION Rather slender, domed tree with dark greyish-brown bark, smooth in young trees but ridged in old ones. Branches are slender and spreading. Leaves are >15cm-long and even more deeply lobed than Pin Oak's (see p. 85), but less strongly bristle tipped. In summer the leaves are glossy green above and paler below, with small hair-tufts in the vein axils; they turn brilliant red in autumn, especially in the cultivar 'Splendens'. Acorns >2.5cm long are rounded and half-enclosed in a slightly glossy cup. **HABITAT** Native to E North America. Occurs in dry, sandy and acidic soils in lowland areas, often in woodland. **DISTRIBUTION** Planted in Europe for its brilliant autumn colours.

Wych Elm ■ *Ulmus glabra* Height >40m

DESCRIPTION Large, often spreading tree, frequently with several prominent trunks arising from stout bole. Smooth greyish bark in younger trees, becomes browner with deep, vertical cracks and ridges. Main branches are spreading, sometimes almost horizontal. Youngest twigs are thick, reddish-brown and covered with short, stiff hairs; older twigs are smoother and greyer. In winter buds are reddish-brown, hairy and oval, with blunt-pointed tips. Leaves are rounded or oval, >18cm long, each with a long, tapering point at the tip. Base of leaf is unequal: a good pointer to all the elms. Leaves feel rough. Flowers are sessile with purple anthers, opening in early spring. Papery fruits are about 2cm long. **HABITAT** Native to much of Europe, including Britain and Ireland. Riversides, valleys, hedgerows and woodland margins, often in wetter areas. **DISTRIBUTION** Occurs in most of Europe, but susceptible to Dutch Elm disease and does not regenerate with suckers.

English Elm ▪ *Ulmus procera* Height >36m

DESCRIPTION Tall, domed tree with dark brown bark that is grooved with small, squarish plates. Main branches are large and ascending. Twigs are thick, reddish and densely hairy.

Winter buds are 3mm long, ovoid, pointed and minutely hairy. Leaves feel rough and are rounded or slightly oval, each with a short, tapering tip; base is unequal, and the longest side does not reach beyond the petiole to twig. Flowers have dark red anthers; they open before the leaves in early spring. Ripe fruits (rarely produced) are >1.5cm long, papery and very short stalked. **HABITAT** Native to S and E Europe, growing in woodland margins, hedgerows and small copses. **DISTRIBUTION** Probably introduced to Britain millennia ago. Formerly widespread and prized for timber, fodder and shelter, but now much reduced by Dutch Elm Disease.

European White Elm
▪ *Ulmus laevis* Height >20m

DESCRIPTION Broadly spreading tree with an open crown and grey bark that is smooth when young, and deeply furrowed with age. Twigs are reddish-brown and softly downy, but become smooth with age. Leaves are >13cm long, with markedly unequal bases and toothed margins. Leaf veins are paired, and longer side has 2–3 more veins than the other. Upper leaf surface is usually smooth, but underside is normally grey-downy. Flowers are produced in long-stalked clusters. Fruits are winged, papery, have a fringe of hairs and grow in pendulous clusters. **HABITAT** Native to mainland Europe, growing in woodland edges, hedgerows and copses, and on hillsides. **DISTRIBUTION** Possibly native to Britain in the past, but now probably extinct. Sometimes grown in collections.

Small-leaved or Field Elm

▪ *Ulmus minor* ssp. *minor* Height >32m

DESCRIPTION Domed and spreading tree with ascending branches bearing pendulous masses of shoots. Leaves resemble Hornbeam's (see p. 76), and are >15cm long, oval, pointed at the tip and have toothed margins; unequal leaf bases, narrowly tapering on short side, and a short petiole. Fruits are papery. Subspecies *minor* includes trees previously (and sometimes still) known as Smooth-leaved Elm *U. carpinifolia* and Coritanian Elm *U. coritana*. **HABITAT** Native to S and SE England, favouring hedgerows, woodland margins and small copses, mostly in lowland areas. **DISTRIBUTION** Once widespread, but range and abundance have been badly affected by Dutch Elm Disease.

Dutch Elm

▪ *Ulmus x hollandica* Height >30m

DESCRIPTION Tall and rather straggly hybrid tree. Higher branches are longer than branches lower down, and are spreading. Bark is brown, cracking into small, shallow plates. Leaves are oval, toothed, >15cm long and sometimes buckled. Leaf base is only slightly unequal. Papery fruits appear in late summer. Naturally occurring hybrid, whose parents are presumed to include Wych, Plot's and Small-leaved Elms. Some natural resistance to Dutch Elm Disease, especially in certain cultivars, notably 'Groeneveld'. **HABITAT** Hedgerows, lowland farmland, woodland margins and copses. **DISTRIBUTION** Scattered range across S England and SW Wales, and found in hedgerows in lowland districts.

Cornish Elm ▪ *Ulmus stricta (U. minor* ssp. *angustifolia)* Height >36m

DESCRIPTION Tall tree with a narrowly conical and rather open crown, and relatively few branches, the lowest ones ascending steeply. Bark is grey-brown and scaly. Leaves are oval, toothed, small (>6cm), and smooth and leathery above, downy on the midrib below. Leaf is narrow, almost equal at the base, and sometimes concave and with a straight midrib. Petiole is 1cm long and downy. Papery fruits appear in late summer. Much reduced because of Dutch Elm Disease. **HABITAT** Hedgerows, farmland, woodland margins and copses. **DISTRIBUTION** Restricted mainly to Cornwall and W Devon; more local elsewhere in West Country and introduced to SW Ireland. Occasionally seen in tree collections.

Plot's Elm

▪ *Ulmus plotii* Height >25m

DESCRIPTION Distinctive tree with a narrow, upright crown, arching leading shoot and rather slender branches with long, pendulous twigs. Bark is greyish-brown and scaly. Leaves are narrower than English Elm's (see p. 87), widest in the middle, and have a straight midrib, pointed tip and almost equal base; upper surface feels rough. Typical elm papery fruits form in summer. **HABITAT** Native to English East Midlands, favouring hedgerows, woodland edges and farmland in lowland areas with damp soils. **DISTRIBUTION** Occasionally planted outside its native range.

Caucasian Elm

■ *Zelkova carpinifolia* Height >31m

DESCRIPTION Tree with a dense, multi-stemmed crown composed of numerous almost upright branches leading to characteristic 'goblet' shape. Bole, >3m, is heavily ridged and has greyish, flaking bark that falls away in rounded scales, exposing orange patches. Leaves are >10cm long, oval and pointed, with rounded teeth and 6–12 pairs of veins. Male flowers, produced in spring, are sessile clusters of yellow-green stamens arising from older, leafless parts of twigs. Female flowers are solitary and in axils of last few leaves on shoot. Fruits are spherical, >5mm across and slightly four winged. Mature trees produce suckers and can spread like hedgerow elms. HABITAT Native to the Caucasus. Grows on hillsides and in woodland, often on moist soils. DISTRIBUTION Grown outside its native area for ornament. Noted for its autumn colours.

Keaki

■ *Zelkova serrata* Height >26m

DESCRIPTION Large tree on a short bole with many spreading branches, giving rise to a 'goblet' shape. Young twigs are hairy at first, becoming smoother with age, and buds grow in a zigzag pattern. Leaves are markedly toothed and smooth below, with even leaf margins. Fruits are smooth and rounded, not papery like those of elms. Tolerant of a range of soil types, and a very long-lived tree. Highly resistant to Dutch Elm Disease. HABITAT Native to Japan and E China. Grows on riversides and in woodland margins, preferring full sun. DISTRIBUTION Widely planted as an ornamental tree.

Black or Common Mulberry
■ *Morus nigra* Height >13m

DESCRIPTION Gnarled bole and dense, twisting branches and twigs make even a young Black Mulberry look ancient. Crown may be broader than tree is tall, and bark is dark orange-brown, fissured and peeling. Downy shoots release milky juice if snapped. Leaves are >20cm long and oval, each with a heart-shaped base, toothed margin and pointed tip. Flower spikes are produced on short, downy stalks in spring. Yellowish-green male flowers are about 2.5cm long. Female flowers are about 1–1.25cm long and give rise to a hard, raspberry-like fruit that is acidic until fully ripened, when it is wine-red or purple. **HABITAT** Native to Asia. Grows on sunny hillsides and in sheltered valleys, often on stony ground. **DISTRIBUTION** Long cultivated in Asia and elsewhere. In Britain found mainly in south, in sheltered gardens.

White Mulberry ■ *Morus alba* Height >15m

DESCRIPTION Deciduous tree with a narrow, rounded crown on a broad bole, >2m across. Heavily ridged grey bark, sometimes tinged pinkish. Shoots are thin, with fine hairs at first; buds are minute, brown and pointed. Leaves are >18cm long, oval to rounded, each with a heart-shaped base and a hairy, grooved petiole >2.5cm long. Leaf feels thin and smooth, and has a toothed margin, with downy hairs on veins on underside. Female flowers are stalked, spike-like and yellowish. Male flowers grow on slightly longer spikes, and are whitish with prominent anthers. Fruit comprises a cluster of drupes that are white or pink at first, ripening to purple. Leaves are foodplant of silkworms. **HABITAT** Native to E Asia. Mostly seen in cultivation in China and neighbouring regions. **DISTRIBUTION** Occasionally grown in Europe, mainly as a specimen tree.

Fig ■ *Ficus carica* Height >5m

DESCRIPTION Deciduous tree with distinctive fruits and leaves. Bark is pale grey, smooth and sometimes with fine lines. Branches are thick, forming a spreading domed crown. Leaves are alternate, >20cm long and grow on a 5–10cm petiole; they are deeply lobed, and usually in three (sometimes five) segments. They feel rough and leathery, and have

prominent veins on undersides. Flowers are hidden, each produced inside a pear-like, fleshy receptacle that is almost closed at the apex. This ripens in the second year into a familiar fleshy, sweet-tasting fig. **HABITAT** Native to SW Asia, and possibly also S and E Europe. Grows on dry, stony ground, on hillsides and in valleys, often in cultivation. **DISTRIBUTION** Long cultivated in Britain, thriving in walled gardens. Several cultivars have been developed, and these are more likely to be seen in gardens. Fruits are eaten either fresh or dried.

Barberry
■ *Berberis vulgaris* Height >2m

DESCRIPTION Small deciduous shrub with grooved twigs and numerous long, three-forked prickles. Attractive flowers are small, yellow and borne in hanging clusters in late spring. Fruits are ovoid reddish berries, borne in loose clusters; they are edible but rather acidic. Leaves are sharp-toothed, oval and arise in tufts from axils of prickles; they produce good autumn colour. **HABITAT** Found in hedgerows and scrub, mainly on calcareous soils. **DISTRIBUTION** Scarce native in Britain, but also planted and naturalized. Poisonous, but has been used as a medicinal herb.

Katsura Tree ▪ *Cercidiphyllum japonicum* Height >25m

DESCRIPTION Conical-crowned deciduous tree, sometimes with a single bole, more often with several main stems. Bark is vertically fissured and peeling. Leaves grow in opposite pairs, and are >8cm long, rounded, and have pointed tips and heart-shaped bases; pink at first, turning green in summer, then red in autumn. Flowers are produced in leaf nodes in spring. Male flowers are small clusters of reddish stamens. Female flowers are darker red clusters of styles. Fruits are claw-like bunches of 5cm-long pods that change from grey, through green, to brown. **HABITAT** Native to Japan, growing on hillsides, and in scrub and woodland, preferring moist soils and high rainfall. **DISTRIBUTION** Grown in Britain and Ireland for ornament, especially for its autumn colours.

Tulip Tree ▪ *Liriodendron tulipifera* Height > 45m

DESCRIPTION Impressive deciduous tree – despite its name, its foliage is more attractive than the flowers. Leaves are strikingly shaped, >20cm long and four lobed with a terminal notch, looking as though they have been cut out with scissors; they are fresh green through summer, and turn bright gold in autumn. Flowers are superficially tulip-like. They are cup shaped at first and greenish, blending in with the leaves. Later they open more fully, revealing rings of yellowish stamens surrounding paler ovaries. They are often produced high up in the middle of dense foliage, and not until the tree is at least 25 years old. Conical fruits, >8.5cm long, are composed of numerous scale-like, overlapping carpels. **HABITAT** Native to E USA, in forests, and on valley floors and mountain slopes, preferring rich soils. **DISTRIBUTION** Introduced into Europe in the 17th century and commonly planted in gardens and parks.

(Southern) Evergreen Magnolia or Bull Bay

▪ *Magnolia grandiflora* Height >30m

DESCRIPTION Large, spreading evergreen tree with a broadly conical crown and large branches, the youngest shoots covered with thick down and ending in red-tipped buds. Leaves are elliptical, >16cm long, and have a smooth or sometimes wavy margin. Upper

surface is shiny dark green, and underside is rust coloured. Flowers are fragrant, composed of six white, petal-like segments and open out to a spreading cup shape >25cm across. Fruit is conical, >6cm long and composed of scale-like carpels on a single orange stalk. **HABITAT** Native to SE USA, favouring lowland swamps, riverbanks and forests, but widely cultivated in parks and gardens. **DISTRIBUTION** Introduced into Europe in the 18th century. Popular in gardens and does well if grown against a wall. In more sheltered areas forms a splendid free-standing tree.

Bay or Sweet Bay

▪ *Laurus nobilis* Height >17m

DISTRIBUTION Moderate-sized evergreen tree with a conical crown and mostly ascending branches, terminating in reddish twigs and conical dark red buds. Bark is smooth and dark grey or almost black. Leaves are glossy, >10cm long, narrowly oval or lanceolate with wavy margins; they feel leathery and have a pleasing aroma when crushed. Flowers are borne below the petioles against the shoots, and open in creamy-yellow clusters. Fruits are small, shiny black berries, >1.5cm long. **HABITAT** Native to Mediterranean area. Grows on hillsides and in woodland margins, often in cultivation. **DISTRIBUTION** Widely planted outside its native region and naturalized in S Britain. Long favoured as a culinary herb.

Californian Laurel
■ *Umbellularia californica* Height >20m

DESCRIPTION Dense evergreen with a domed crown and similar foliage to Bay's (see p. 94). Branches are much divided and bark is grey and cracked. Leaves are similar to Bay's, but typically narrower and paler green or yellowish-green. Crushed leaves give off a stronger scent than those of Bay, which can induce headaches and nausea in some people. Yellowish flowers grow in small, dense, rounded clusters, and fruits are rounded and greenish, ripening to purple. **HABITAT** Native to W North American coast, favouring coastal hills and grassland, woodland and mountain slopes. **DISTRIBUTION** Cultivated in mild parts of Europe, where some fine trees exist.

Sassafras ■ *Sassafras albidum* Height >20m

DESCRIPTION Medium-sized, columnar deciduous tree with thick, reddish-brown, furrowed and aromatic bark. Branches bear thin green shoots. Leaves are mostly elliptic, >15cm long and 10cm across, sometimes with large lobes on either side. Upper leaf surface is bright green, lower surface bluish-green; leaves turn through yellow and orange, to purple in autumn. Can form large clumps that produce a brilliant autumn display. Crushed leaves have a pleasing smell and taste of orange and vanilla. Male and female flowers are very small, greenish-yellow and grow in small clusters on separate plants. Fruit is an ovoid berry, about 1cm long, ripening to dark blue. **HABITAT** Common native tree of E North America, used as a raw ingredient for root beer and tea. Grows in woods and thickets, and also in cultivation. **DISTRIBUTION** Seen in Britain and Ireland in arboreta and well-established gardens.

Sweet Gum ■ *Liquidamber styraciflua* Height >28m

DESCRIPTION Large tree with attractive foliage. Leaves are sharply lobed with toothed margins. They are alternate and give off a resinous scent when crushed, unlike maple

leaves, which they resemble. Branches are twisting and spreading to upcurved, and bark is greyish-brown. Flowers are globose; fruits are spiny and pendulous, 2.5–4cm across, resembling fruits of London Plane (see p. 99). **HABITAT** Widespread and common native tree of SE USA as far south as Central America. Occurs mainly in lowland areas on good soils, but sometimes on mountain slopes and hillsides. **DISTRIBUTION** Familiar in Britain as a colourful autumn tree in many parks and gardens. Also imported as timber called satinwood.

Persian Ironwood ■ *Parrotia persica* Height >12m

DESCRIPTION Small, spreading deciduous tree with a short bole. Smooth bark, peeling away in flakes, and leaving attractive coloured patches; older trees have a pattern of pink, brown and yellow. Branches are mostly level, and young twigs are hairy, terminating in

blackish, hairy buds. Leaves are >7.5cm long and oval, each with a slightly tapering tip and rounded base; they become very colourful in autumn. Flowers appear before leaves in short-stalked clusters, and are reddish and inconspicuous. **HABITAT** Native to Caucasus and N Iran, favouring hillsides, mountain slopes and woodland margins. **DISTRIBUTION** Introduced into Europe as an ornamental tree for its outstanding autumn colours.

Witch Hazel

■ *Hamamelis mollis* Height >4m

DISTRIBUTION Rarely more than a small, sprawling shrub, but sometimes grows into a small, domed tree. Leaves resemble Hazel's (see p. 77), and are alternate and mostly oval, each with a pointed tip, toothed margin and unequal base. Best known for conspicuous winter flowers, produced long before the leaves open and providing a welcome sign of early spring. They are composed of long yellow, ribbon-like petals and red stamens, and are noticeably sweet scented. **HABITAT** Native to China. Grows in woodland margins, clearings and gardens, and on hillsides. **DISTRIBUTION** Introduced into Britain late in the 19th century. Now found in parks and gardens, and sometimes naturalized in open woodland.

Japanese Witch Hazel ■ *Hamamelis japonica* Height >4m

DESCRIPTION Spring-flowering shrub or small tree with a spreading habit, rather subdued flower colour and colourful autumn leaves. Leaves are large and similar to Hazel's (see p. 77), opening long after the flowers. Flowers are produced in late winter or very early in spring and have a stronger scent than those of Witch Hazel (see above). **HABITAT** Native to Japan. Favours hillsides, mountain slopes, woodland margins and clearings, and often grown in parks and gardens. **DISTRIBUTION** Introduced into Europe as an ornamental tree, mainly for its autumn colours.

Kohuhu ■ *Pittosporum tenuifolium* Height >10m

DESCRIPTION Stout-boled tree with smooth, dark grey bark and densely packed branches ending in purplish-black shoots. Leaves are oblong or elliptical, >6cm long and 2cm across, with a wavy margin; they are glossy above and less shiny below. Delicately scented, tubular flowers, >1cm long, have five deep-purplish lobes and yellow anthers; they appear in clusters or singly, in leaf axils. Fruit is a rounded capsule, about 1cm long, ripening from green to black. **HABITAT** Native to New Zealand, growing in coastal areas and sheltered lowland sites on a variety of soils. **DISTRIBUTION** Planted in Europe but not hardy, thriving only in milder coastal areas in south-west, or in very sheltered gardens.

Karo ■ *Pittosporum crassifolium* Height >10m

DESCRIPTION Small evergreen tree or large shrub with blackish bark and numerous congested thin branches. Leaves are leathery, >8cm long and 3cm wide, ovate to lanceolate and blunt-tipped; dark green above, paler and woolly below, with slightly

inrolled margins. Flowers appear in lax clusters, and have five deep red petals and yellow anthers. Fruit is an ovoid capsule, >3cm long, with shiny seeds. **HABITAT** Native to New Zealand, growing in coastal areas, on cliffs and hillsides. **DISTRIBUTION** Tolerant of salt spray, so planted for coastal hedging and naturalized in parts of SW England; also planted as a garden shrub in sheltered areas. Foliage is popular with florists.

Oriental Plane ■ *Platanus orientalis* Height >30m

DESCRIPTION Large deciduous tree with a broad, domed crown. Main trunk is frequently covered with large, tuberous burrs and smooth bark flaking away to leave yellow patches.

Leaves are large, >18cm long and wide; each is deeply divided into 5–7 lobes and borne on a 5cm-long petiole with a swollen base enclosing a bud. Female flowers, >8cm long, comprise up to six rounded, dark red flowerheads. As they ripen into fruits, the ball-like heads grow >3cm across; they contain many one-seeded carpels with long hairs attached to the bases. **HABITAT** Native to Balkans, eastwards into Asia. Grows on hillsides and mountain slopes, and in forests and river valleys. **DISTRIBUTION** Commonly planted in British parks and gardens, sometimes alongside roads.

London Plane ■ *Platanus x hispanica (x acerifolia)* Height >44m

DESCRIPTION Very large deciduous tree resulting from cross between **American Plane** *Platanus occidentalis* and Oriental Plane (see above), known since the mid-17th century.

Main trunk is usually very tall and the crown of an old tree is often spreading. Greyish-brown bark flakes away in rounded patches, leaving paler, yellowish areas. Branches are often tangled and twisted. Leaves are variable, >24cm long, mostly five lobed and palmate, but not as deeply divided as those of Oriental Plane. Fruits are similar to Oriental Plane's. **HABITAT** Most common in urban situations, but also found in parks and gardens. **DISTRIBUTION** Widespread in towns, where the peeling bark provides a useful way of ridding the tree of dust deposits. Extremely sturdy and resistant to gales, storms and disease. Very much a feature of London streets and squares, but also popular in other cities.

Oleaster or Silver Berry ■ *Elaeagnus angustifolia* Height >13m

DESCRIPTION Deciduous shrub or small tree with grey bark and spiny branches, and silvery young twigs. Leaves are lanceolate, >8cm long and 2.5cm wide; they are silvery below and dull green above. Flowers are bell shaped, >1cm long, and grow singly or in pairs in leaf axils on 2–3mm pedicels; they are silvery outside, yellow inside. Fruits are oval, >1cm long and yellow with silvery scales. **HABITAT** Native to W Asia. Grows in dry, rocky areas and scrub, on hillsides and in valleys, often on poor soils. **DISTRIBUTION** Occasionally planted in Europe as an ornamental tree; naturalized in some areas.

Bridewort

■ *Spiraea salicifolia* Height >2m

DESCRIPTION Much-divided shrub that sends up suckers freely, helping it to spread rapidly. Whip-like branches are mainly upright and have reddish-brown bark. Leaves are elliptical to narrowly oval, with the margin toothed towards the tip. Small flowers have tiny pink petals; they grow in numerous dense, conical to cylindrical heads, creating a colourful spectacle in summer. **HABITAT** Native to central and E Europe, growing in boggy areas, forest margins and stream-sides, and preferring acid soils. **DISTRIBUTION** Popularly grown in British gardens; occasionally naturalized on roadsides and in hedgerows.

Quince
▪ *Cydonia oblonga* Height >7.5m

DESCRIPTION Small, irregularly spreading deciduous tree with a flattened crown, dense branches and shoots that are noticeably woolly at first, but lose this trait as they become older. Leaves are green and smooth above, and grey and downy below. Attractive pink-tinged white flowers, which are at their best in spring, are >5cm in diameter and bowl shaped. Fruit is >3.5cm long; it resembles a small pear, and is greenish at first and becomes golden-yellow when mature, with a pleasing fragrance. **HABITAT** Native to SW Asia, favouring hillsides and woodland, but often grown in cultivation. **DISTRIBUTION** Long cultivated outside its native region, including in our area, for its fruits. Mostly found in orchards and gardens, but occasionally naturalized in hedges and open woodland.

Willow-leaved Pear ▪ *Pyrus salicifolia* Height >10m

DESCRIPTION Small deciduous tree with a rounded crown and mostly level branches

bearing pendulous, very downy twigs. Bark is rough, scaly and usually dark brown. Leaves resemble willow leaves: they are silvery-grey on both surfaces at first, but greener on the upper surface later in the season. White flowers, >2cm across, usually open at the same time as leaves. Fruit is about 3cm long, pear shaped or sometimes more pointed, and brown when ripe. **HABITAT** Native to central Asia. Grows on open, scrubby hillsides, and in woodland margins and hedgerows. **DISTRIBUTION** Grown in Europe as an ornamental tree, mainly for its foliage.

Plymouth Pear ■ *Pyrus cordata* Height >8m

DESCRIPTION Small, slender or slightly spreading deciduous tree with spiny branches and purplish twigs. Dark brown bark breaks up into small, square plates. Leaves are

alternate, oval and >5cm long, but often much smaller. Leaf is downy when young, becoming dull green when older. Flowers open at the same time as leaves, and the tree is often covered with white blossom. Fruit resembles a tiny pear on a long stalk; it is golden-brown at first, ripening to red and marked by numerous brown lenticels. Suckering is an important means by which Plymouth Pear reproduces. **HABITAT** Native to SW Britain, W France and Iberian peninsula, growing in hedgerows, copses and woodland margins. **DISTRIBUTION** Scarce in its range.

Wild Pear ■ *Pyrus pyraster* Height >20m

DESCRIPTION Medium-sized deciduous tree, sometimes becoming fairly large and spreading. Dark brown bark breaks up into small, square plates. Branches are usually spiny.

Leaves are elliptic, each >7cm long and with a partly toothed margin. Flowers open at the same time as leaves. Petals, sometimes looking slightly crushed, are borne in clusters of about five. Tree is often densely covered with blossom. Small, hard fruits form in autumn. **HABITAT** Native to Europe, including S Britain, growing in open woodland, copses and hedgerows. Usually solitary, but typically easy to spot for a short time in spring when the white blossom is open. **DISTRIBUTION** Grows in a wide area of Europe.

Common or Cultivated Pear ■ *Pyrus communis* Height >20m

DESCRIPTION Slender, upright deciduous tree with a stout bole and dark brown bark breaking up into small, square plates. Some branches may bear a few spines. Young twigs

are reddish-brown and sparsely hairy, becoming smoother with age. Leaves are usually oval to elliptic, smooth and almost glossy when mature. Pure white flowers open before leaves. Fruits are >12cm long, with soft, slightly gritty, sweet-tasting flesh. **HABITAT** Originally native to W Asia. Mostly known from cultivation, but occurs in woods and hedgerows as an escape. **DISTRIBUTION** Cultivated for millennia and now widespread across Europe, including Britain and Ireland.

Oleaster-leaved Pear ■ *Pyrus elaeagrifolia* Height >10m

DESCRIPTION Small, often slender deciduous tree with rough, scaly bark and spreading, spiny branches bearing twigs covered with grey hairs. Leaves are lanceolate, >8cm long

and covered with thick white down, even at the end of the growing season. White, almost sessile flowers open with leaves. Thick-stalked fruits are about 1.3cm long and pear shaped, sometimes globular, remaining green when ripe. **HABITAT** Native to Balkans eastwards. Grows on stony hillsides, and in scrub, woodland margins and cultivated areas. **DISTRIBUTION** Occasionally planted outside its native area as an ornamental tree because of its compact shape.

Sage-leaved Pear
■ *Pyrus salvifolia* Height >10m

DESCRIPTION Small, much-branched tree with rough, scaly bark. Spreading branches are spiny with blackish, almost hairless old twigs. Leaves are >5cm long, elliptical, and smooth above, and grey and woolly below. White flowers open with leaves, followed by pear-shaped fruit, >8cm long. Pedicel and young fruits are woolly; the bitter fruit ripens to yellow and is used to make Perry. Possibly a hybrid between Common and Snow Pears (see pp. 103 and 105). HABITAT Occurs in the wild from France eastwards. Grows in hedgerows, open areas, woodland margins and clearings. DISTRIBUTION Occasionally planted outside its native region.

Almond-leaved Pear ■ *Pyrus amygdaliformis* Height >6m

DESCRIPTION Small, hardy tree with rough, scaly bark, dense and sparsely spiny branches, and greyish, woolly young twigs. Leaves are >8cm long, each with a sparsely toothed margin. Young leaves are downy, and full-grown leaves are shiny above and slightly downy below. White flower clusters open with leaves. Thick-stalked fruits are rounded, >3cm across, ripening to dark yellow, but remaining rather hard. HABITAT Native to SE Europe, favouring dry, open areas, hillsides, woodland margins and scrub, and tolerant of poor soils. DISTRIBUTION Planted outside its native area as an ornamental tree; fruit is not considered edible.

Snow Pear

■ *Pyrus nivalis* Height >20m

DESCRIPTION Medium-sized tree with rough, scaly bark and ascending, usually spineless branches. Leaves are >9cm long and smooth; the blade runs decurrently down the petiole. From a distance the tree looks silvery, especially in windy weather. Showy pure white flowers open just after leaves. Fruits are >5cm long, rounded and greenish-yellow with purple dots; considered edible either raw or cooked. **HABITAT** Native in area from France to Russia. Favours hedgerows, woodland margins and clearings, hillsides and scrub. Tolerant of dry conditions. **DISTRIBUTION** Planted outside its native area as an ornamental tree for its spring blossom, fruits and autumn colour.

Wild Crab

■ *Malus sylvestris* Height >10m

DESCRIPTION Small tree, sometimes tall and slender if growing in woodland, or more spreading if occurring in a hedgerow. Bark is deep brown, cracking into small plates. Branches are numerous and may be spiny, and smooth brown shoots often have spines. Leaves are >11cm long, toothed and smooth surfaced. Flowers are >4cm across and usually white, although they may show a pinkish tinge. Cultivated varieties of apple, which have become naturalized, always show pink tinge. Fruits are >4cm in diameter, rounded, yellowish-green when ripe, and tough and sour to taste. **HABITAT** Native to Europe, including Britain and Ireland, except far north. Woodland, hedgerows and scrub, often on heavy soils. **DISTRIBUTION** Occurs in wide area of Europe.

Plum-leaved Crab
■ *Malus prunifolia* Height >10m

DESCRIPTION Upright tree that becomes rather straggly when neglected or naturalized, with dense and irregularly divided branches. Greyish-brown bark develops striking fissures in mature trees. Leaves are broadly oval and plum-like, >7cm long, and shiny above but rather hairy below when young. Flowers are >5cm across, pinkish, and grow in dense clusters among terminal leaves on shoots.
HABITAT Native to China. Mostly seen in parks and gardens, but tolerant of a range of soils and habitats >1,300m.
DISTRIBUTION Planted rather occasionally in parks and gardens in Britain and Ireland; very occasionally naturalized.

Japanese Crab
■ *Malus floribunda* Height >8m

DESCRIPTION Compact, densely crowned, small tree on a thick bole, with dark brown, fissured bark. Twigs are slightly pendulous and reddish when young, and remain densely hairy. Alternate leaves are >8cm long and oval, each with a pointed tip and toothed margin; underside is downy at first, but smooth later. Fragrant flowers appear soon after leaves and are usually so dense that they hide the leaves. Buds are pink at first, but flowers fade to white. Fruits are rounded and >2.5cm across, ripening to bright yellow, and often appear in the same abundance as the flowers. **HABITAT** Probably a hybrid between two Japanese garden species, so has not been found growing in the wild.
DISTRIBUTION Frequently planted in Europe for its attractive blossom and convenient small size.

Siberian Crab ▪ *Malus baccata* Height >15m

DESCRIPTION Domed and rather spreading deciduous tree with dense, much-divided branches and brownish, cracking bark. Leaves are rather slender, and matt rather than shiny. Compact flowerheads

of white blossom make this a popular garden tree. Fruit is green at first, but ripens to bright red and remains on tree long after leaves have fallen, providing a valuable late feed for winter migrant birds. Hardy and disease resistant. **HABITAT** Native to China, growing in mixed forests and on mountain slopes >1,500m; tolerant of various soil types. **DISTRIBUTION** Planted in Britain and Ireland in parks and gardens.

Cultivated Apple ▪ *Malus domestica* Height >15m

DESCRIPTION Familiar orchard tree producing copious quantities of edible fruits. Bark is usually brown and branches can be untidy unless pruned. Leaves are elliptical and rounded at the bases, each with a slightly pointed tip and toothed margin, and slightly downy above and more downy below. Flowers are white or tinged with pink and, in some varieties, produced abundantly in short-stalked clusters. Fruits are normally >5cm in diameter, and indented at the pedicel. A great variety of shapes, sizes, tastes and colours exists. **HABITAT** Almost always found in cultivation in orchards and gardens across much of Europe. **DISTRIBUTION** Occasionally naturalized, or found in isolated places where human habitation once occurred or where apple cores, containing seeds (pips) have been discarded. Cultivated Apple is a hybrid species, probably between Wild Crab (see p. 105) and **Paradise Apple** M. *dasyphylla*.

Hawthorn-leaved Crab
■ *Malus florentina* Height >10m

DESCRIPTION Attractive and rather conical small fruit tree with good blossom, small fruits and colourful autumn foliage. Bark is brown with yellowish scales, and the branches are mostly level to upright. Leaves are >8cm long and sharply lobed, resembling those of Wild Service-tree (see p. 110). Flowers are whitish, and the small, rounded fruits, around 1cm across, ripen to red. Possibly a hybrid between another *Malus* species and Wild Service-tree. HABITAT Parks, gardens, woodland margins and clearings. Mostly known from cultivation. DISTRIBUTION Planted in parks and gardens for its compact growth habit and autumn colour.

Rowan or Mountain Ash ■ *Sorbus aucuparia* Height >20m

DESCRIPTION Small to medium deciduous tree with a fairly open, domed crown, ascending branches, and purple-tinged twigs that are hairy when young, becoming smooth later. Bark is silvery-grey and smooth. Leaves are composed of 5–8 pairs of toothed leaflets. Flowers are produced in dense heads in spring; each is >1cm in diameter with five creamy-white petals. Fruits are rounded, <1cm long and bright scarlet, hanging in colourful clusters and persisting after leaves have fallen. HABITAT Native to Europe. Hardy tree found in woodland and open country, and on hillsides and mountains. DISTRIBUTION Found in a wide area of Europe, and often planted as a town tree in squares and along roadsides.

True Service-tree ■ *Sorbus domestica* Height >20m

DESCRIPTION Resembles Rowan (see p. 108), but note subtle differences in bark, buds and fruits. Rich brown bark is fissured, ridged and often peels in vertical shreds. Buds are smooth, rounded and green, unlike the purple, pointed buds of Rowan. Leaves are alternate and pinnate, composed of up to eight pairs of oblong, toothed leaflets about 5cm long, and softly hairy on undersides. Flowers are produced in spring in rounded, branched clusters; each flower is about 1.5cm across and composed of five creamy-white petals. Small, pear- or sometimes apple-shaped fruits are >2cm long, and green or brown. They have a very sharp taste when ripe, but after a frost become more palatable. **HABITAT** Native to many parts of Europe, North Africa and south-west Asia; rare British native. Favours hillsides, woodland margins, cliffs and scree. **DISTRIBUTION** Widespread but generally rare; planted occasionally in tree collections.

Sargent's Rowan ■ *Sorbus sargentina* Height >10m

DESCRIPTION Small, often much-branched tree with smooth bark and twigs. Sticky red winter buds open to produce pinnate leaves with 4–5 pairs of 5cm-long, sharply toothed leaflets, hairy below. They turn a striking red colour in autumn. Flowers are small and white, produced in dense heads in late spring. Fruits are bright red clusters of berries often produced prolifically. **HABITAT** Discovered in the early 20th century in W China by American botanist Charles Sargent. Favours open, sunny hillsides and woodland glades and margins, usually on well-drained soils. **DISTRIBUTION** Popular in Europe in parks and gardens for its fiery red autumn colours.

Wild Service-tree ■ *Sorbus torminalis* Height >25m

DESCRIPTION Medium-sized deciduous tree with spreading or sometimes domed habit. Finely fissured bark breaks into squarish brown plates, thought to be the origin of one of its vernacular English names, Chequers. Twigs are shiny and dark brown, terminating

in rounded shiny green buds. Leaves are >10cm long, each with 3–5 pairs of pointed lobes and sharply toothed margins; basal lobes project at right angles, while other lobes point forwards. White flowers are >1.5cm in diameter, and grow in loose clusters on woolly pedicels. Fruit is >1.8cm in diameter, rounded and brownish, with numerous lenticels in the skin. **HABITAT** Native to much of Europe, including Britain, and Asia Minor and North Africa. Favours copses, ancient woodland, hedgerows and cliffs, often on heavy soils. **DISTRIBUTION** Relatively rare, and sometimes planted for its good autumn colours.

Common Whitebeam
■ *Sorbus aria* Height >25m

DESCRIPTION Small to medium deciduous tree, with a spreading or domed crown. Smooth grey bark is sometimes ridged. Twigs are brown on upper surfaces and usually green below, and hairy when young. Buds are >2cm long, ovoid and green, tipped with white hairs. Leaves are simple and oval, and very hairy, especially on the white undersides. When they first open their underside colour can make a distant tree appear white. White flowers are produced in stalked clusters, opening in spring. Fruits are ovoid, about 1.5cm long, bright red by autumn and covered with many small lenticels. **HABITAT** Native to S Britain and a large part of mainland Europe. Most common on chalk and limestone, usually in open, sunny sites and woodland margins, and on hillsides. **DISTRIBUTION** Widely planted as an ornamental tree.

French Hales or Devon Whitebeam

■ *Sorbus devoniensis* Height >7m

DESCRIPTION Medium-sized tree, or sometimes large hedgerow shrub, with a broadly conical shape. Leaves are oval and leathery with shallow-toothed, sharp lobes on distal two-thirds of leaf; they are dark glossy green above and white below, with 7–9 pairs of veins. White flowers appear in clusters in spring and edible fruits, ripening in late summer, are brownish orange, >15mm long, with numerous lenticels. **HABITAT** Endemic to Devon and E Cornwall, and SE Ireland. Favours hedgerows, woodland margins, copses and steep hillsides, on neutral to slightly acid, rocky ground. **DISTRIBUTION** In the wild, local in its native range.

Swedish Whitebeam ■ *Sorbus intermedia* Height >15m

DESCRIPTION Medium-sized deciduous tree with a compact, tidy shape when mature. Leaves are >12cm long, oval and deeply lobed. They are glossy green above, but yellowish and downy below. Clusters

of small white flowers appear in late spring. They are followed by bunches of scarlet fruits that are oval, >1.5cm long and pitted with many lenticels. They remain on the tree after the leaves have fallen, providing winter food for birds. **HABITAT** Native to Scandinavia, favouring open areas, scrub, hillsides, woodland margins and cliffs. **DISTRIBUTION** Hardy and tolerates air pollution, so popular in urban situations.

Hupeh or Hubei Rowan ■ *Sorbus hupehensis (S. glabrescens)* Height >15m

DESCRIPTION Small, dome-shaped deciduous tree with slender branches and shoots, and grey-brown to purple, smooth bark. Leaves are pinnate with 11–17 pairs of narrow, oval

leaflets with serrated margins. They change from green to orange or red in autumn. Clusters of white flowers appear in spring, followed by numerous small, rounded, pink-tinged fruits in autumn. **HABITAT** Native to China. Favours hillsides, woodland margins and clearings, and tolerant of a range of soil conditions. **DISTRIBUTION** Very popular outside its native range in gardens because of its growth habit, autumn leaf colour and contrasting fruits.

Snowy Mespil
■ *Amelanchier ovalis* Height >5m

DESCRIPTION Small deciduous tree or shrub. Leaves are >5cm long, with coarsely toothed margins and downy undersides when first open. Flowers appear in late spring on upright spikes of up to eight white-petalled flowers with yellowish anthers; these are followed by pea-sized, edible red fruits that ripen to become blue-black in autumn. **HABITAT** Native to mainland Europe eastwards. Favours open, sunny and usually dry hillsides, forest margins and clearings, and tolerant of many soil types. **DISTRIBUTION** Planted in gardens outside its native range as an ornamental shrub; occasionally naturalized in Britain.

American Snowy Mespil

■ *Amelancheir laevis* Height >10m

DESCRIPTION Small, smooth-barked trees with similar leaves to Snowy Mespil (see p. 112), which have a bronze tinge at first then turn bright red in autumn. Pure white, star-shaped flowers appear on drooping spikes in spring with leaves, and can be very showy in mature trees. Later they give rise to numerous 6mm-long, edible fruits, which are purple when ripe and very attractive to garden birds. **HABITAT** Native to USA, growing in woodland clearings, hedgerows, hillsides and scrub, often on sandy or acid soils. **DISTRIBUTION** Introduced into Europe and naturalized in parts of S England, but usually seen in parks and gardens.

Canadian Snowy Mespil ■ *Amelancheir canadensis* Height >10m

DESCRIPTION Small tree, or often a large shrub with several stems and a narrow crown. Leaves are alternate, oval, downy on undersides and >5cm long; each leaf has a serrated margin and pointed tip. Flowers are produced in spring in loose clusters of 4–10; the five petals are pure white, surrounding cream-coloured stamens. Edible rounded fruits appear in late summer, turning purple when fully ripe. **HABITAT** Native to NE America, growing in woodland, scrub, hillsides and valleys, and preferring damp soils and wet lowland habitats. **DISTRIBUTION** Planted outside its native range as an ornamental garden tree; occasionally naturalized in woodland.

Juneberry ■ *Amelanchier grandiflora (lamarckii)* Height >9m

DESCRIPTION Small deciduous spreading tree, or sometimes large shrub, with hairy young twigs. Leaves are >7cm long and elliptical, with finely toothed margins, purple tinged and woolly at first, but becoming smooth and green when older. Flowers grow in drooping, slightly hairy spikes; petals are white, >1.8cm long. Small, rounded edible fruits are deep purple when ripe, with dried sepals persisting at the tips; their flavour is similar to that of some apples. **HABITAT** Open, sunny areas on well-drained soils, mainly in lowland regions. Probably a hybrid between two similar wild species, first arising in North America. **DISTRIBUTION** Planted in Britain and mainland Europe as a garden shrub; sometimes naturalized.

Himalayan Tree Cotoneaster ■ *Cotoneaster frigidus* Height >20m

DESCRIPTION Medium-sized tree or many-stemmed shrub with pale grey bark and slender, spineless branches. Leaves are elliptical with entire margins, dark green above, and hairy and white below. They persist in most winters and are semi-evergreen. Flowers are small, and grouped into dense white clusters about 5cm across. Numerous small fruits form in autumn, >5mm long, becoming round and red. They are a good source of food for birds in winter. **HABITAT** Native to the Himalayas. Grows on hillsides, and in scrub and woodland margins, on a variety of soil types. **DISTRIBUTION** Widely planted outside its native range as an ornamental shrub; sometimes naturalized.

Medlar

■ *Mespilus germanica* Height >9m

DESCRIPTION Sometimes a small, rounded tree, or often a spreading and untidy shrub. Bark is greyish-brown, in old trees breaking into oblong plates. Young shoots are densely hairy. Leaves are >15cm long, often a yellowish-green colour and almost shiny above, with dense white hairs on undersides. Solitary white flowers are >6cm across, with sepals longer than petals and about 40 red anthers. The curious fruit is about 3cm long, with a brown, russet-like skin and sunken apex. It is edible, but not until it has started to rot, when it can be used in preserves. **HABITAT** Native to SE Europe and Asia Minor, growing in woodland and scrub in sunny areas on dry soils. **DISTRIBUTION** Long cultivated for its edible fruits; sometimes naturalized.

Cockspurthorn

■ *Crataegus crus-galli* Height >10m

DESCRIPTION Small, usually spreading deciduous tree, with a flattish crown and short bole. Bark is smooth and greyish-brown in young trees; in older trees it is fissured. Purple-brown twigs carry numerous 7–10cm-long, sharp spines. Leaves are >8cm long and about 3cm wide, with toothed margins. Both surfaces are smooth and shiny, turning a rich orange in early autumn. White flowers are about 1.5cm in diameter and grow in loose clusters, opening in spring. Red globular fruits ripen in autumn and persist after leaves have fallen. **HABITAT** Native to NE America, growing in hedgerows, copses, woodland margins and scrub. **DISTRIBUTION** Often planted in our area as a garden or roadside tree, mostly for its striking autumn colours.

Hybrid Cockspurthorn ■ *Crataegus x lavallei* Height >12m

DESCRIPTION Dense, domed to spreading tree with grey, scaly bark. Branches are level, with thorn-bearing twigs growing thickly on upper sides, a feature making winter

recognition easy. Leaves are narrow and glossy green, turning dark red late in autumn but persisting for some months. Flowers are white and borne in clusters; fruits are dull red, >18mm long and also persist in winter. **HABITAT** Hedgerows, scrub, woodland margins and gardens. **DISTRIBUTION** Common in town gardens and on roadsides. Tolerant of mild pollution and coastal conditions, so popular for urban situations.

Broadleaf Cockspurthorn ■ *Crataegus persimilis* Height >8m

DESCRIPTION Small, spreading tree with a short bole and rounded crown. Bark is scaly brown, sometimes with spiral ridges. Twigs and 2cm-long, stiff thorns are deep, glossy purple-brown. Leaves are oval and toothed, smooth and shiny; they turn from glossy green, through yellow, orange and copper, to deep red by the end of autumn before they finally fall. Flowers are white and borne in clusters; fruits are rounded and red, >1.5cm long. **HABITAT** North American species. Grows in hedgerows, scrub, woodland margins and roadsides. **DISTRIBUTION** Popularly planted in our area in roadsides and gardens, typically as the cultivar 'Prunifolia'; also makes a good stock-proof hedge.

Pear-fruited Cockspurthorn ■ *Crataegus pedicellata* Height >7m

DESCRIPTION Small tree with a rounded crown and spreading branches; twigs bear numerous long thorns. Leaves are broadly ovate, lobed and double toothed, with a slightly glossy surface, turning violet. Flowers are white with a pink tinge, and appear in loose clusters in spring. In autumn pear-shaped, glossy-skinned red fruits, >2cm long, are formed, and these may persist through winter. **HABITAT** Native to NE USA, growing in hedgerows, scrub, woodland margins and clearings, and avoiding wet soils. **DISTRIBUTION** Introduced to Europe, and now often planted and naturalized in many areas.

Common Hawthorn
■ *Crataegus monogyna* Height >15m

DESCRIPTION Small, spreading deciduous tree or hedgerow shrub. Bark is usually heavily fissured into a pattern of vertical grooves. Branches are densely packed, along with twigs, and there are numerous sharp spines. Leaves are >4.5cm long, roughly ovate and deeply lobed. Flowers are produced in flat-topped clusters of about 10–18; they are white or sometimes pink tinged. Each flower is about 1.5cm in diameter with a single style in the centre. Fruits (haws) are usually rounded and bright red, or sometimes darkening to maroon; they contain a hard-cased seed. **HABITAT** Native to much of Europe, favouring woodland, hedgerows, hillsides, scrub, cliffs and wasteground. **DISTRIBUTION** Very common across most of Europe, and widely planted as a stock-proof hedgerow species.

Oriental Hawthorn

■ *Crataegus laciniata* Height >10m

DESCRIPTION Rather low and spreading tree, with scaly brown bark showing pinkish patches. Branches are often twisted; young twigs and pedicels are covered with white hairs; becoming smooth and blackish with age. Leaves are deeply lobed and >4cm long, with fine white hairs on both sides. Flowers are creamy-white, and occur in dense clusters of >16. Fruits are hairy at first; they ripen to orange or red, and contain 3–5 seeds. **HABITAT** Native to SE Europe, Spain and Sicily. Favours hedgerows, scrub, woodland margins and clearings, and tolerant of a wide range of soils. **DISTRIBUTION** Occasionally planted in our area for ornament.

Midland Hawthorn

■ *Crataegus laevigata* Height >10m

DESCRIPTION Small, dense tree or large shrub. Grey-brown bark cracks into regular-shaped plates, revealing darker patches beneath. There are some short spines in the twig axils. Leaves are >6cm long, but not as deeply lobed as those of Common Hawthorn (see p. 117). Lobes are rounded and toothed to the bases. Flowers are borne in rather lax clusters of up to nine normally white flowers that are each >2.4cm in diameter. Fruits are about 1cm long, rounded and deep red in colour. **HABITAT** Native to central and W Europe, including England, favouring woodland, hedgerows, hillsides and scrub, often on heavy soils. **DISTRIBUTION** Widespread but not as frequent as Common Hawthorn.

Peach ▪ *Prunus persica* Height >6m

DESCRIPTION Small, bushy, rounded deciduous tree with dark brown bark. Branches are straight, with smooth, reddish, angular twigs. Leaves are alternate, lanceolate, finely toothed and often creased into a V shape. Pink flowers, >4cm across, are usually solitary; they open at the same time as leaf buds and have yellow-tipped anthers. Fruit is the familiar peach, >8cm long, rounded, downy and flushed pink; it has sweet, juicy flesh when ripe. Seed is contained inside a woody, thickly ridged 'stone'. **HABITAT** Probably native to China, but long cultivated elsewhere and known mainly from cultivation. Thrives in sheltered places protected from strong winds and frosts. **DISTRIBUTION** In our area does best in walled gardens.

Almond ▪ *Prunus dulcis* Height >8m

DESCRIPTION Small, open-crowned tree whose blossom appears early in spring. Branches are ascending, usually spiny and with numerous thin twigs, but many cultivars are regularly branched and lack spines. Leaves are alternate, >13cm long, finely toothed and folded lengthways. Cup-shaped pink or white flowers are paired and almost sessile, each with five petals >2.5cm long. Fruit is about 6cm long, flattened ovoid, and covered with velvety-green down with a tough, fleshy layer below; inside there is a ridged and pitted 'stone' that when cracked reveals the edible almond seed. **HABITAT** Probably native to central and SW Asia, and N Africa, but long cultivated for seeds and flowers. Favours dry, sunny sites in sheltered areas. **DISTRIBUTION** In our region requires protection from harsh winter weather.

Apricot ■ *Prunus armeniaca* Height >10m

DESCRIPTION Small, rounded deciduous tree, with greyish-brown bark with fine fissures. Twisted and dense branches bear smooth reddish twigs. Leaves are heart shaped, and reddish when first open, later becoming green above and yellowish beneath, on a red petiole with two glands near the leaf base. White or pale pink, short-stalked flowers, solitary or paired, open before leaves. Fruit, >8cm long, is rounded, the downy red-tinged skin surrounding a rather acid-tasting, juicy flesh that becomes sweet only when fully ripe. Stone is flattened, elliptical and smooth, with three raised lines along one edge. **HABITAT** Native to central and E Asia. Grows in sunny, sheltered areas on a range of soils. **DISTRIBUTION** Grown for its edible fruits; requires shelter in our region.

Cherry Plum or Myrobalan Plum ■ *Prunus cerasifera* Height >8m

DESCRIPTION Small deciduous, rather bushy tree. Dark brown bark is pitted with rows of white lenticels, and in older trees is fissured. Many fine and sometimes spiny branches and glossy green twigs. Leaves are >7cm long and ovate, with toothed margins. White, stalked flowers are mostly solitary and open at about the same time as leaves. Fruits, >3.5cm long, are rounded, with smooth red or yellow skin; flesh becomes sweet and stone inside is rounded but has a thickened margin. **HABITAT** Native to Balkans. Grows in hedgerows, scrub, roadsides, woodland margins and clearings. **DISTRIBUTION** Widely planted in our area for its edible fruits; frequently naturalized.

Blackthorn or Sloe

■ *Prunus spinosa* Height >6m

DESCRIPTION Rather untidy-looking, densely branched, spiny deciduous tree with very dark bark. Branches are spreading, terminating in spiny twigs. Leaves are oval, 4.5cm long, and each pointed at the tip and with toothed margins. White flowers open before leaves; they are often produced so prolifically that whole hedgerows can turn white in early spring. Fruits (sloes) are >1.5cm long, rounded to slightly ovoid, and blue-black with a whitish bloom on the thin skin. The flesh inside is very acidic, and stone is mostly smooth. **HABITAT** Native to much of Europe apart from far north, growing in hedgerows, scrub, roadsides and woodland margins, and often forming dense thickets. **DISTRIBUTION** Widespread and common. Often planted as a stock-proof hedge.

Common Plum ■ *Prunus domestica* ssp. *domestica* Height >10m

DESCRIPTION Small deciduous tree with dull brown bark, sometimes tinged purple, with deep fissures developing with age. Branches are usually straight with no spines; twigs are brown and smooth. Leaves are >8cm long, with toothed margins, smooth green upper surfaces and downy lower surfaces. Flowers are mostly white, hanging in small clusters of 2–3, and opening at about the same time as leaves. Fruits are >7.5cm long, rounded or more often oval, with a smooth skin. Probably a hybrid between Blackthorn and Cherry Plum (see above and p. 120). **HABITAT** Mostly seen in cultivation in gardens and orchards in lowland areas. **DISTRIBUTION** Widely planted throughout Britain and Ireland; occasionally naturalized, nearly always near human habitation.

Bullace

■ *Prunus domestica* ssp. *instita* Height >10m

DESCRIPTION Small deciduous tree with dull brown bark. Branches are usually spiny, and young twigs and flower stalks are downy. Leaves are >8cm long with toothed margins, and green above and downy below. Flowers are mostly white or sometimes green tinged, and grow in small clusters of 2–3, opening at about the same time as leaves. Rounded fruits are about 5cm long, with purplish skin. Flesh is rather acidic at first, but ripens to become sweet and juicy. Oval stone is flattened, and usually rough and slightly pitted. Probably a hybrid between Blackthorn and Cherry Plum (see pp. 121 and 120). **HABITAT** Hedgerows, scrub, woodland margins and roadsides. **DISTRIBUTION** Widely planted in Britain and Ireland; also naturalized, nearly always near human habitation.

Dwarf or Sour Cherry

■ *Prunus cerasus* Height >8m

DESCRIPTION Small deciduous tree with a very short, branching bole and rounded, shrubby outline, often surrounded by suckers. Leaves are >8cm long, oval and sharply pointed, with toothed margins. Long-stalked white flowers usually open just before leaves, growing in clusters of 2–6. Fruits >1.8cm long are rounded with a slightly depressed apex, and usually bright red or blackish-red. Flesh is soft and tastes acidic, and stone is rounded and smooth. **HABITAT** Native to SW Asia. Favours hedgerows, scrub, roadsides, woodland clearings and margins. **DISTRIBUTION** Widely cultivated for its edible fruit. Planted in our region; widely naturalized, usually by suckering and not bird sown.

Wild Cherry or Gean
■ *Prunus avium* Height >30m

DESCRIPTION Large deciduous tree with a tapering bole and high, domed crown. Reddish-brown bark is shiny, with circular lines of lenticels, peeling horizontally into tough, papery strips. Branches spread widely, terminating in smooth, reddish twigs. Leaves are >15cm long, ovate and with a pointed apex and forwards-pointing teeth. White flowers grow in long-stalked clusters of 2–6. Fruits are >2cm long, rounded and dark purple, red-black or sometimes yellow, with shiny skin. Flesh may be sweet tasting or rather bitter. Stone is rounded and smooth. **HABITAT** Native to much of Europe except far north and east, favouring woodland, copses, hillsides and hedgerows. **DISTRIBUTION** Widespread, and frequently planted and naturalized.

Sargent's Cherry ■ *Prunus sargentii* Height >13m

DESCRIPTION Open, spreading tree with purple-brown, rather glossy bark ringed with horizontal bands. Branches are ascending or spreading, and the dark red twigs are thin and

smooth. Leaves are >15cm long and ovate, each with a pointed tip and toothed margins, and smooth on both surfaces. Pale pink flowers grow in clusters of 2–4, opening just before leaves. Petals are >2cm long. Fruits (rarely seen here) are ovoid, >1.1cm long and dark crimson. Usually grafted on to a stock of Wild Cherry (see above). **HABITAT** Native to Japan and Sakhalin Islands. Grows in woodland clearings, on hillsides, and in parks, gardens and urban sites. **DISTRIBUTION** Widely planted in our region.

Rum or Black Cherry
■ *Prunus serotina* Height >22m

DESCRIPTION Spreading deciduous tree with a stout trunk and greyish bark that peels away in strips and is fissured in older trees; a strange, bitter smell is released if bark is damaged. Branches are spreading and dense. Leaves are >14cm long, shiny above and with fine, forwards-pointing teeth on margins; midrib on leaf underside has patches of hairs along it, unlike in other similar cherries. Flowers are similar to Bird Cherry's (see p. 127), but spike may contain fewer than 30 flowers. Black fruits contain bitter-tasting flesh and a rounded, smooth stone. **HABITAT** Native to North America, favouring hedgerows, wasteland, woodland margins and scrub. **DISTRIBUTION** Planted for timber and ornament in much of Europe, including Britain and Ireland; naturalized in many places.

Japanese Cherry ■ *Prunus serrulata* Height >15m

DESCRIPTION Small to medium deciduous tree, noted for its shiny, purple-brown bark ringed by horizontal lines of prominent lenticels. Branches are ascending, usually fanning

out from the bole and ending in smooth twigs. Leaves are >20cm long and ovate, with toothed margins and a long, tapering tip. White or pink flowers grow in clusters of 2–4, opening just before leaves. Fruits are round, >7mm long and deep purple-crimson; they seldom develop in cultivated trees. **HABITAT** Probably native to China. Occurs in woodland and on hillsides, but mostly seen in cultivation. **DISTRIBUTION** Introduced into Japan at a very early date; subsequently brought to our region, where it is now a very popular garden tree.

Saint Lucie Cherry
■ *Prunus mahaleb* Height >12m

DESCRIPTION Often little more than a spreading shrub, but sometimes a small tree. Branches are spreading, and young twigs are covered with short greyish hairs and often slightly weeping at the tips. Leaves are >7cm long, almost rounded, and have a short point at the tip and finely toothed margins. White, scented flowers grow in clusters of 3–10 in groups of racemes at ends of leafy shoots. Fruit is a 0.6–1cm-long, rounded black berry with bitter-tasting flesh. HABITAT Native to central and S Europe. Grows in woodland glades, thickets and scrub. DISTRIBUTION Planted in our region for ornament; occasionally naturalized.

Spring Cherry ■ *Prunus subhirtella* Height >20m

DESCRIPTION Densely crowned deciduous tree with greyish-brown bark and slender branches, and many downy, crimson twigs. Leaves are >6cm long, ovate to lanceolate, and have a long-pointed tip and irregularly toothed margins; the veins are downy below.

Pinkish-white, short-stalked flowers open just before leaves in spring; petals are about 1cm long and notched. Purplish-black fruits are rounded but seldom produced. Various cultivars have been developed, some with a weeping habit, others with double flowers; they are usually grafted on to stocks of Wild Cherry. HABITAT Native to Japan. Grows on hillsides, and in woodland glades and margins, cultivated areas and urban sites. DISTRIBUTION Commonly planted as a street and garden tree in Britain and Ireland.

Winter Cherry ■ *Prunus x subhirtella* 'Autumnalis Rosea' Height >8m

DESCRIPTION Small deciduous tree or large shrub with a spreading habit if not crowded by other trees. In many respects similar to Spring Cherry (see p. 125), but flowers open very early, often in late autumn or early winter, and persist in mild winters. Leaves are

ovate, >6cm long, and green above but paler and downy below; they produce a fine autumn colour. Dark, rounded, shiny fruits may sometimes be produced. **HABITAT** Rarely seen outside parks and gardens. Grows best on well-drained soils in sunny sites. **DISTRIBUTION** Widely planted as a garden tree for its winter blossom and autumn leaf colour.

Tibetan Cherry

■ *Prunus serrula* Height >15m

DESCRIPTION Small deciduous tree resembling Spring Cherry (see p. 125). Deep purple bark peels to reveal a rich and glossy, mahogany-coloured inner layer, which is often rubbed smooth by passers-by. Branches spread widely and are often pruned to reveal more of the bark, the main decorative feature of this tree. Leaves are >12cm long, lanceolate and pointed. White flowers open with leaves. Bright red fruits about 1cm long sometimes form. **HABITAT** Native to China, growing on hillsides, and in woodland margins and clearings, often in cultivation. **DISTRIBUTION** Occasionally planted in our region for its attractive bark, seen at its best in winter.

Bird Cherry ■ *Prunus padus* Height >17m

DESCRIPTION Deciduous tree with smooth, grey-brown bark that releases a strong, unpleasant smell if rubbed. Thin, ascending branches end in twigs that are smooth, but finely downy when young. Leaves are >10cm long, and have dark green upper surfaces and slightly blue-green undersides, and finely toothed margins. White flowers open after leaves, grow in 15cm-long spikes and smell of almonds. Fruits are >8mm long, shiny black and sour tasting, rather like sloes. HABITAT Native to Europe, including Britain and Ireland. Favours stream-sides, damp woods, cliffs and hedgerows. DISTRIBUTION Widespread; common in limestone areas, and sometimes planted.

Portuguese Laurel
■ *Prunus lusitanica* Height >8m

DESCRIPTION Small, spreading evergreen tree, or usually a shrub, with widely spreading branches, very dark bark and smooth, reddish twigs. Leaves are lanceolate, dark green, glossy, slightly leathery and >13cm long. White flowers are borne in long, tapering spikes of about 100 strongly scented flowers. Fruits are >1.3cm long, ovoid or rounded, with a tapering tip; they are purplish-black when ripe and contain a smooth, rounded stone with a ridged margin. HABITAT Native to Portugal, Spain and SW France. Favours hillsides, woodland margins and clearings, abandoned gardens and cliffs. DISTRIBUTION Frequently planted outside its native region for ornament; can survive regular clipping. Now widely naturalized, spreading by layering and by self-sown seeds.

Cherry Laurel

■ *Prunus laurocerasus* Height >8m

DESCRIPTION Evergreen shrub or small, spreading tree with dense branches. Bears smooth, pale green twigs and leathery leaves >20cm long and 6cm wide, each with a short-pointed tip and smooth margins with just a few very small teeth. Fragrant white flowers are borne in an erect spike about 13cm long. Fruits are rounded and green at first, turning red, then blackish-purple. **HABITAT** Native to E Balkans, favouring hillsides, scrub, woodland margins and clearings. **DISTRIBUTION** Commonly planted as an ornamental since the 16th century in S and W Europe. In Britain and Ireland often naturalized, and also exists in a number of cultivars.

Judas Tree ■ *Cercis siliquastrum* Height >10m

DESCRIPTION Small, spreading and rather flat-crowned deciduous tree, often with more than one bole. Branches are ascending, spreading near tips, with red-brown buds and twigs. Leaves are simple, alternate and rounded, and sometimes notched at the tip and heart shaped at the base; they are smooth above and bluish-green when young, becoming yellow when older. Five-petalled, pink, pea-like flowers grow in small short-stalked clusters; they

open before the leaves and burst out of the bole, large branches and twigs. They are followed by pods, >10cm long, which are slightly constricted around the seeds, reddish at first and maturing to brown, and becoming dehiscent. **HABITAT** Native to E Mediterranean. Favours stony hillsides and open, sunny sites, and often planted in towns and gardens. **DISTRIBUTION** Planted outside its native region for ornament.

Honey Locust ■ *Gleditsia triacanthos* Height >45m

DESCRIPTION Tall deciduous tree with a high, domed crown; the bole, branches and twigs are spiny, and the bark is greyish-purple. Branches are mainly level with the curled twigs. Leaves are either pinnate, with up to 18 pairs of 2–3cm-long leaflets, or bi-pinnate, with up to 14 leaflets. Tiny, 3mm-long flowers open in June; they may be male, female or both, and grow in compact clusters in leaf axils; greenish-white, oval petals number 3–5. Flattened pods with thickened edges, >45cm long, are twisted and become dark brown when ripe. **HABITAT** Native to Mississippi basin of North America, in riversides, valleys, woodland margins and clearings. **DISTRIBUTION** Planted in our region for ornament.

Laburnum ■ *Laburnum anagyroides* Height >7m

DESCRIPTION Deciduous tree with a narrow, sparse crown, and slender bole with smooth, greenish-brown bark marked with blemishes. Branches are often slightly pendulous; shoots are grey-green with long, silky, clinging hairs. Leaves are divided into three, each leaflet >8cm long, elliptic and blunt pointed at the tip; leaves are hairy below when young. Yellow, fragrant, pea-like flowers appear copiously in 10–30cm-long pendulous racemes in early summer. Pods, >6cm long, have smooth, blackish-brown, dry outer skin. They persist on the tree, twisting open to reveal pale inner skin and dark seeds, which are very poisonous. **HABITAT** Native to central and S Europe, growing on hillsides, and in woodland and cultivated areas. **DISTRIBUTION** Planted in our region for ornament; sometimes naturalised.

False Acacia ■ *Robinia pseudoacacia* Height >30m

DESCRIPTION Medium-sized, open-crowned tree with spirally ridged bark and branches that snap easily. Leaves are alternate, >20cm long and pinnate, with 3–10 pairs of oval, yellowish-green leaflets; petiole has two woody basal stipules, and each leaflet has a small stipule at the petiole base. Fragrant white, pea-like flowers, >20cm long, grow in dense hanging clusters, and are produced prolifically in early summer. Pods are smooth, >10cm long. **HABITAT** Native to USA. Favours hillsides, woodland margins and clearings, and roadsides, on a variety of soil types. **DISTRIBUTION** Widely planted in our region for ornament; frequently naturalized.

Tree of Heaven
■ *Ailanthus altissima* Height >20m

DESCRIPTION Vigorous, suckering tree with smooth grey bark at first, but becoming pale and scaly with age. Branches are thick and mostly upright; twigs end in tiny, ovoid scarlet buds. Leaves are alternate, pinnate and >60cm long, with >25, 7–12cm-long, pointed leaflets; they are deep red when first emerged, but shiny green in summer. Greenish flowers grow on fairly open spikes on separate sex trees. Fruits are reddish and winged, and the twisted seeds are about 3cm long. **HABITAT** Native to China. Grows on hillsides and in woodland margins, preferring limestone soils but tolerant of most soil types. **DISTRIBUTION** Widely planted outside its native region as an ornamental; also naturalized.

Stag's-horn Sumach ■ *Rhus typhina* Height >10m

DESCRIPTION Small, spreading tree with brown bark and downy, spreading branches.
Leaves are alternate, pinnate and have >29 leaflets; each leaflet is >12cm long and coarsely
toothed. Leaflets are green at
first, but noted for their fiery
autumn colours. Tiny flowers
are borne on separate-sex
trees; greenish male flowers
and red female flowers grow
in dense, conical clusters,
>20cm long, at tips of twigs.
Fruits resemble small nuts.
HABITAT Native to N
America. Grows in open
areas, woodland margins and
clearings, and scrub patches.
DISTRIBUTION Widely
planted in our region as
an ornamental; sometimes
naturalized and rather
invasive.

Field Maple
■ *Acer campestre* Height >26m

DESCRIPTION Medium-sized deciduous
tree with a rounded crown and twisted bole.
Variable in appearance, depending on habitat.
Grey-brown bark has a slightly corky texture.
Branches are much-divided and dense. Shoots
are brown, sometimes covered with fine
hairs and often developing wings, especially
on trees that are regularly pruned back in
hedgerows. Leaves are >12cm long and three
lobed. Newly opened leaves have a pinkish
tinge, become dark green and leathery later,
and turn yellow in autumn. Yellowish-green
flowers are borne in small, erect clusters.
Winged fruits grow in bunches of four.
HABITAT Native to N Europe. Favours
woods and copses, hedgerows, hillsides and
cliffs, on a variety of soils. **DISTRIBUTION**
Widespread and common.

Sycamore

■ *Acer pseudoplatanus* Height >35m

DESCRIPTION Fast-growing deciduous tree with a spreading habit and domed crown. Greyish bark is broken up by numerous fissures into irregular patches. Branches are usually quite thick near main bole, and end in grey-green twigs with reddish buds. Leaves are >15cm long and divided into five toothed lobes. Flowers are slender, pendulous, yellow clusters >12cm long. Paired, winged fruits ripen in late summer and reach a length of 6cm. HABITAT Native to hills and uplands of central and S Europe. Favours woodland, hillsides, hedgerows, sea cliffs and urban sites. DISTRIBUTION Widely planted and naturalized outside its native region, doing well on heavy soils, and tolerates coastal conditions as well as uplands.

Norway Maple

■ *Acer platanoides* Height >30m

DESCRIPTION Tall, spreading deciduous tree with a short bole, high, domed crown and thinner branches than those of Sycamore (see above); twigs are green, sometimes tinged red. Smooth grey bark is sometimes slightly ridged. Bright green leaves are smooth, each >15cm long with 5–7 toothed, pointed lobes; lowest pair of lobes is smaller than others. Greenish-yellow flowers, in compact erect clusters of 30–40, open before leaves; males and females are separate. Paired, yellowish fruits are >10cm across. HABITAT Native to mainland Europe. Grows on hillsides and cliffs, and in woodland and urban sites, on a range of soil types. DISTRIBUTION Widely planted in Britain and Ireland for ornament; also naturalized. Cultivar A. *platanoides* 'Crimson King' with dark red foliage is used in municipal plantings.

Montpelier Maple

■ *Acer monspessulanum* Height >15m

DESCRIPTION Small deciduous tree with a neatly domed crown, and branches with thin brown twigs terminating in small, orange-brown buds. Bark is dark and fissured. Leaves are leathery and >8cm long; each leaf has three spreading lobes and entire margins, and is shiny dark green above and bluish below with a few tufts of hairs in the axils of the lower veins. Leaves are fresh green in spring, but dark in summer, and remain on tree until well into autumn. Yellowish-green flowers open after leaves, in small clusters on long, slender pedicels. Red-tinged fruits are about 1.2cm long, with parallel or overlapping wings. **HABITAT** Native to S Europe. Favours hillsides, woodland margins and clearings, and cultivated areas. **DISTRIBUTION** Planted outside its native region for ornament.

Ashleaf Maple or Box Elder ■ *Acer negundo* Height >20m

DESCRIPTION Small, vigorous deciduous tree with numerous shoots growing from the bole and main branches, giving it an untidy appearance. Bark is smooth in young trees, becoming darker and shallowly fissured when older. Branches have green shoots and small buds. Leaves are pinnate, >15cm long, with up to seven irregularly toothed, oval leaflets. Male and female flowers occur separately, opening before leaves. Flowers lack petals; male flowers are greenish with prominent red anthers, female greenish-yellow and pendent. Brown fruits are about 2cm long with slightly spreading wings. **HABITAT** Native to E North America. Grows in woodland, scrub, floodplains and riversides. **DISTRIBUTION** Commonly planted for ornament, and sometimes for shelter; sometimes naturalized.

Smooth Japanese Maple ■ *Acer palmatum* Height >16m

DESCRIPTION Small deciduous tree with a short, twisted bole, domed crown and smooth brown bark with paler patches in young trees. Spreading branches end in thin, reddish twigs with green undersides. Leaves are >11cm long, with toothed lobes divided at least

halfway to bases. Dark purple-red flowers grow in upright clusters of 12–15 on thin, 4cm-long pedicels. Reddish fruits usually hang in clusters; each fruit is about 2cm across, with wings diverging widely. The cultivar 'Atropurpureum' has fine purple leaves through the season; 'Osakazuki' has brilliant scarlet autumn leaves and fruits. **HABITAT** Native to Japan, favouring hillsides, woodland clearings and scrub patches. **DISTRIBUTION** Planted in our region for its compact shape, interesting foliage and fine autumn colours. Cultivars 'Atropurpureum' and 'Osakazuki' are commonly planted in parks and gardens.

Downy Japanese Maple ■ *Acer japonicum* Height >14m

DESCRIPTION Similar to Smooth Japanese Maple (see above), but the bole is often even shorter. Bark is grey and smooth, and branches are upright and sinuous. Leaves are hairy

when young, with veins remaining hairy through the season. Leaves are lobed, divided less than halfway to bases, with forwards-pointing teeth. Purple flowers grow in long-stalked, pendulous clusters, opening just before leaves. Paired winged fruits, >5cm across, have wings diverging widely; margins are hairy at first. The cultivar 'Vitifolium' has red autumn colours; 'Aconitifolium' has deeply divided leaves. **HABITAT** Native to Japan. Grows on hillsides and slopes, and in woodland margins and glades. **DISTRIBUTION** Grown in our region for ornament.

Sugar Maple ■ *Acer saccharum* Height >26m

DESCRIPTION Large deciduous tree similar to Norway Maple (see p. 132). Bark has large fissures and falls away in shreds in older trees. Branches are upright to spreading. Leaves are 13cm-long and lobed, and teeth on lobes are rounded, not drawn out into a fine point as in Norway Maple; there are hairs in vein axils below. Leaves produce fine autumn colours. Pendulous yellow-green flowers are small, lack petals and open in spring with leaves. Winged fruits appear in autumn. **HABITAT** Native to E North America, favouring forests, mountain slopes and valley floors, often in mixed woodland. **DISTRIBUTION** Widely planted outside its native region for its autumn colours.

Silver Maple ■ *Acer saccharinum* Height >30m

DESCRIPTION Broadly columnar deciduous tree with a spreading crown; suckers freely. Bark is smooth and greyish, becoming scaly with age. Numerous slender, ascending branches bear pendulous brownish twigs. Leaves are >16cm long, deeply divided into five lobes and have irregularly toothed margins. They are orange or red tinted at first and green above later, but with silvery hairs below; petiole is usually pink tinged. Yellowish-green flowers appear in small, short-stalked clusters of separate sexes in spring. Green, then brown fruits are about 6cm long, with diverging wings and prominent veins. **HABITAT** Native to E North America. Grows in riversides, valleys, marshes and wet ground, and also in urban sites. **DISTRIBUTION** Planted in our region for ornament. Cultivar 'Laciniatum' is common in city squares.

Red Maple ■ *Acer rubrum* Height >23m

DESCRIPTION Fast-growing, spreading tree with an irregular crown and mostly ascending branches that arch outwards. Bark is grey and smooth. Leaves are >10cm long and almost

as wide, each with 3–5 toothed lobes less than half the leaf width. They are red tinged above at first, greener later, and silvery below, with a red petiole, turning various shades of red and yellow in autumn. Small red flowers in dense clusters open in spring before leaves. Bright red, winged fruits are about 1cm long, the wings diverging at a narrow angle. **HABITAT** Native to E North America, usually growing in damp sites such as those on riverbanks, and in marshes and lowland areas. **DISTRIBUTION** Grown in Europe for its colourful autumn foliage.

Nikko Maple ■ *Acer nikoense* Height >15m (rarely 20m)

DESCRIPTION Broadly spreading deciduous tree with greyish-brown, smooth bark. Branches are mainly level, with blackish buds that have grey hairs on scales. Leaves are compound, each with three leaflets, the central one >10cm long, the other two smaller

and unequal at the base. They are green and smooth on the upper surface, bluish-white below with a covering of soft hairs; they turn fiery red in autumn. Small yellow flowers, in pendulous clusters of three, open at about the same time as leaves. Green, winged fruits are about 5cm long, and wings spread widely. **HABITAT** Native to Japan. Grows in forests and on hillsides, and often in cultivation. **DISTRIBUTION** Now popular outside its native region as an ornamental, mostly for its fine autumn colours.

Paper-bark Maple
■ *Acer griseum* Height >15m

DESCRIPTION Dense and spreading deciduous tree noted for its distinctive reddish-brown bark, which peels off in thin, papery scales. Branches are mainly level and young shoots are downy. Leaves are pinnate and divided into three blunt-toothed leaflets, each toothed and lobed. Yellow-green flowers are small, growing in drooping clusters. Pale green, winged fruits are about 3cm long; seeds are usually infertile. **HABITAT** Native to China. Grows in upland areas, on mountain slopes and in forests on a range of soil types. **DISTRIBUTION** Occasionally planted outside its native region as an ornamental, mainly for the fine autumn leaf colour and peeling bark.

Moosewood
■ *Acer pennsylvanicum* Height >14m

DESCRIPTION One of the 'snakebark maples' with green bark, vertically striped with reddish-brown or white, becoming greyer with age. Branches are mainly upright. Leaves are >15cm long and about the same width, each with three triangular, tapering, forwards-pointing lobes; central lobe is longest. In summer the leaves are rich yellow-green, with a smooth upper surface and hairy lower surface when first open. In autumn they turn deep yellow. Small, yellow-green flowers appear in spring with leaves. Greenish fruits are about 2.5cm long and have down-curved wings. **HABITAT** Native to North America. Grows on wooded slopes and in upland forests, preferring high humidity and damp soils. **DISTRIBUTION** Planted in our region for its autumn colours.

Père David's Maple ■ *Acer davidii* Height >16m

DESCRIPTION Spreading, open deciduous tree, with bark patterned with green-and-brown, vertical stripes when mature; young twigs have red bark. Branches are mainly upright at first. Leaves are >15cm long, either unlobed and ovate or with shallow lobes;

they are dark green above, paler below, grow on red petioles and produce striking autumn colours. Flowers are yellowish, in 6cm-long drooping racemes. Fruits have wings at a shallow angle. **HABITAT** Native to China. Grows in woods and forests, and on hillsides, preferring shelter and tolerating most soil types. **DISTRIBUTION** Often grown in our region for its ornamental bark, compact shape and autumn colours.

Horse-chestnut ■ *Aesculus hippocastanum* Height >25m

DESCRIPTION Large tree with a massive, domed crown. Greyish-brown bark often flakes away in large scales. Branches snap off readily when large; reddish-brown twigs have numerous whitish lenticels. Winter buds are a conspicuous feature, being shiny brown,

sticky and >3.5cm long. Below each bud is a horseshoe-shaped leaf scar. Leaves are long-stalked and palmate, with up to seven leaflets, each >25cm long, all of them sharply toothed. Flowers often cover tree in a mass of creamy-white panicles, each one made up of 40 or more five-petalled, pink-spotted white flowers. Fruits are spiny cased and rounded, containing a single large, round seed (conker). **HABITAT** Native to mountains of Balkans. Grows on mountain slopes and hillsides, and in forests, often in cultivation. **DISTRIBUTION** Widely planted over much of Europe.

Red Horse-chestnut

■ *Aesculus* x *carnea* Height >20m

DESCRIPTION Hybrid between Horse-chestnut and Red Buckeye (see p. 138 and below), which grows to form a sizeable, domed tree with a gnarled bole and twisted branches. Palmate leaves are composed of 5–7 leaflets, each dark green and with toothed margins. Flowers are similar to those of Horse-chestnut: sometimes creamy-white with yellow blotches at first, but turning pink with red blotches. Seeds are produced in rounded cases, but some cultivars are sterile and do not set seed. **HABITAT** Parks, gardens and urban sites. **DISTRIBUTION** First found in Germany in 1820 and now frequently planted as an ornamental tree, mainly for its red flowers.

Red Buckeye ■ *Aesculus pavia* Height >5m

DESCRIPTION Small, spreading deciduous tree with a domed crown, smooth, dark grey bark, and level or slightly weeping branches. Leaves are palmate, each composed of five

lanceolate, pointed, sharply toothed, short-stalked leaflets; they are dark glossy green above, turning red in autumn. Slender red flowers, >4cm long, have four petals growing in erect spikes in early summer. Fruits are rounded or pear shaped, with a smooth brown outer skin enclosing one or two shiny brown seeds. **HABITAT** Native to SE USA. Grows in understory of mixed woodland, mainly in lowland areas, parks and gardens. **DISTRIBUTION** Planted outside its native region for ornament.

Indian Horse-chestnut ■ *Aesculus indica* Height >30m

DESCRIPTION Large, broadly columnar tree with a thick trunk. Resembles Horse-chestnut (see p. 138), but is more graceful, especially in its winter outline. Bark is smooth greyish-green or pink tinged. Branches are ascending, but with pendulous shoots. Leaves

resemble Horse-chestnut's, but leaflets are narrower, stalked, finely toothed and >25cm long; they are bronze tinged when young, green in summer, and turn yellow or orange in autumn. Flowers are white or pale pink with bright yellow blotches, and stamens extending out of the flower; yellow blotches turn red as flowers matures. Flower spikes are erect and >30cm long. Stalked brown fruits are pear shaped and scaly, with up to three seeds. **HABITAT** Native to the Himalayas. Favours shady valleys and deciduous forests on lower mountain slopes. **DISTRIBUTION** Occasionally planted in our region as an ornamental.

Yellow or Sweet Buckeye
■ *Aesculus flava* Height >30m

DESCRIPTION Large, domed deciduous tree with upright, twisted branches, and peeling and scaly, grey-brown bark. Palmate leaves have five leaflets, each >20cm long. Leaves turn red in early autumn. Four-petalled yellow flowers are borne in erect spikes about 15cm long, usually opening in late spring or early summer. Smooth, rounded fruits are about 6cm across, covered in brown scales on the outside and contain 1–2 seeds. **HABITAT** Native to E USA. Grows in river valleys, floodplains, lowland forests and parks. **DISTRIBUTION** Planted in our region in parks and gardens for its excellent autumn colours.

Holly ■ *Ilex aquifolium* Height >15m

DESCRIPTION Striking evergreen tree with fine, shiny, dark green foliage that has very strong prickles, although leaves higher up may have smooth margins. Sometimes only a shrub, but can grow into a tall, conical tree. With age fissures and tubercles appear on the initially smooth, silver-grey bark. Branches sweep downwards, but tips of younger branches turn up. Shoots and buds are green. White flowers are about 6mm across, four petalled and grow in leaf axils. Males and females grow on different trees. Fruit is a bright red, stalked berry >12mm long. **HABITAT** Woodland, copses, hedgerows, scrub and gardens. **DISTRIBUTION** Native to most of W and S Europe, and parts of W Asia. Widely planted as an ornamental and hedging shrub.

Spindle
■ *Euonymus europaeus* Height >6m

DESCRIPTION Slender, sometimes spreading and rather twiggy deciduous tree with green twigs that are angular when young, but become rounded when older, terminating in tiny, pointed buds. Smooth grey bark becomes slightly fissured and pink tinged as tree ages. Leaves are ovate, >10cm long, with a pointed tip and sharply toothed margins; they turn purple-orange in autumn. Yellowish-green, four-petalled flowers are small, and appear in clusters in leaf axils. Fruits are pink capsules about 1.5cm across and divided into four chambers, each containing an orange seed. **HABITAT** Native to much of Europe except extreme north and south. Grows in hedgerows and copses, especially on lime-rich soils. **DISTRIBUTION** Found throughout the region, and often planted as an ornamental.

Box ■ *Buxus sempervirens* Height >6m

DESCRIPTION Small, very dense, spreading evergreen tree or large shrub with numerous branches; young twigs are green, angular and covered with white hairs. Smooth grey bark breaks into small squares with age. Leaves are opposite, ovate to oblong, >2.5cm long and 1cm across, with a notched tip; upper surface is dark green and glossy, lower surface is paler. Flowers are small and green; male flowers have yellow anthers. Fruit is a small woody capsule with three spreading spines; it splits open to scatter shiny black seeds. **HABITAT** Native to mainland Europe and S England, growing on dry calcareous hillsides and in open areas. **DISTRIBUTION** Rare tree, locally abundant where found. Widely planted for hedging and topiary.

Buckthorn ■ *Rhamnus cathartica* Height >10m

DESCRIPTION Spreading deciduous tree with slender and slightly spiny shoots. Dark orange-brown bark becomes almost black in older trees, but still revealing orange patches between fissures. Leaves are ovate and >6cm long, each with a short, pointed tip and

finely toothed margin. Leaf is glossy green above with a pale underside, and veins on upper surface converge towards leaf tip. In autumn the leaves turn yellow. Fragrant flowers are very small with four, or rarely five, green petals; male and female grow on separate trees in small, stalked clusters, or sometimes singly. Berry-like fruit is black and about 8mm in diameter. **HABITAT** Native to much of Europe; absent from Scotland. Grows in open woods and copses, especially on drier calcareous soils. **DISTRIBUTION** Occurs in most of Europe.

Alder Buckthorn
■ *Frangula alnus* Height >5m

DESCRIPTION Small tree with a spreading habit and smooth, grey, vertically furrowed bark. Twigs have numerous small, fine hairs, are green at first, becoming grey-brown later, and are opposite, like the branches. Leaves are opposite, each >7cm long, and broadly ovate with entire margins and a short-pointed tip; they are glossy green above and paler below, and turn clear lemon-yellow in autumn. Greenish-white, five-petalled flowers are inconspicuous, about 3mm across; they grow in small axillary clusters, opening in summer. Berry-like fruits are black and >10mm in diameter. **HABITAT** Native to much of Europe, apart from far north and drier parts of Mediterranean region, growing in marshy woodland and scrub, mostly on acid soils. **DISTRIBUTION** Occurs throughout England and Wales, but rare and scattered in Scotland and Ireland.

Silver-lime
■ *Tilia tomentosa* Height >30m

DESCRIPTION Broadly domed tree with mostly straight, ascending branches. Young twigs are whitish and woolly, darkening with age; buds are greenish-brown and >8mm long. Bark is grey and ridged. Leaves are >12cm long and rounded, each with a heart-shaped base, tapering tip and toothed margins. They are dark green, hairless and wrinkled above, white and downy with stellate hairs below. Five to ten off-white, strongly scented flowers are supported by a yellowish bract. Fruit, >1.2cm long, is ovoid, warty and downy. Woolly leaves deter aphids, so honeydew – sugary water that aphids extract from some trees – is avoided. **HABITAT** Native from Balkans eastwards. Grows on hillsides, mountain slopes and urban sites. **DISTRIBUTION** Planted in our region and thrives in towns.

Pendent Silver-lime
▪ *Tilia tomentosa* 'Petiolaris' Height >30m

DESCRIPTION Similar deciduous tree to Silver-lime (see p. 143), but the branches have noticeably pendulous tips. Leaf underside is very white and downy, as is the long petiole. When the wind ruffles the leaves the tree looks silvery. Flowers are similar to Silver-lime's but this hybrid species is usually sterile and does not produce viable fruits. May be slightly toxic to some species of bee. **HABITAT** May have originated in Balkans. Most often seen in urban sites, where it is tolerant of drought, pollution and soil compaction. **DISTRIBUTION** Widely planted as a street and park tree.

Small-leaved Lime
▪ *Tilia cordata* Height >32m

DESCRIPTION Tall deciduous tree with a dense crown. Young trees have a neat shape; older trees are less tidy, with burrs, sprouts and criss-crossed branches. Bark becomes flaky and cracked in older trees. Twigs are smooth and brownish-red above, olive below, and buds are dark red. Leaves are >9cm long and rounded, each with a pointed tip, heart-shaped base and finely toothed margins. Flowers are white or pale yellow, fragrant and grow in clusters of >10 on a 10cm-long green bract. Fruit is rounded and hard, and about 6mm in diameter. **HABITAT** Native to Britain, mainland Europe and W Asia. Favours woodland, hillsides and copses, mainly on base-rich soils. **DISTRIBUTION** Widely planted in towns and used for forestry.

Large-leaved Lime
■ *Tilia platyphyllos* Height >40m

DESCRIPTION Tall deciduous tree. Bole normally lacks suckers and shoots, distinguishing it from Lime (see below). Branches are mostly upright with slightly pendent tips. Twigs are reddish-green, and sometimes slightly downy at tips; ovoid buds, >6mm long, are dark red and sometimes slightly downy. Leaves are >15cm long and broadly ovate, each with a short, tapering point and irregularly heart-shaped base. Margins are sharply toothed, and upper surface is soft and dark green, lower surface paler; sometimes slightly hairy. Yellowish-white flowers are borne in clusters of up to six on whitish-green, slightly downy bracts. Hard fruit is >1.8cm long and slightly pear shaped.
HABITAT Native to lime-rich soils in Europe, favouring woodland, hillsides and valleys.
DISTRIBUTION In Britain native to central and S England, and Wales; introduced elsewhere and often planted as a street tree.

Lime ■ *Tilia x europaea (Tilia x vulgaris)* Height >46m

DESCRIPTION Large, upright tree with an irregular crown; bole often has burrs and masses of sprouts, and bark is grey-brown and ridged. Branches are upright, but arching on older trees; young twigs are smooth and green. Buds are 7mm long and reddish-brown.

Leaves are >10cm long and broadly ovate, each with a short, pointed tip, heart-shaped base and toothed margin; they are dull green above, paler below, with tufts of white hairs in vein axils. Yellowish-white, fragrant flowers grow in clusters of up to ten on a greenish-yellow bract. Fruit is hard, thick-shelled and rounded; usually sterile. Hybrid between Small-leaved and Large-leaved Limes (see p. 144 and above). **HABITAT** Woodland and hillsides, often in urban situations. **DISTRIBUTION** Planted and very common in towns and parks, but heavy aphid infestation causing honeydew to rain down renders it unpopular for street planting.

American Lime or Basswood ■ *Tilia americana* Height >25m

DESCRIPTION Broadly columnar deciduous tree with upright branches and greyish-brown bark. Leaves are 20cm long and 15cm wide, and heart shaped with an unequal base, with toothed margins. They are deep green above, paler and more glossy below, with brown hair tufts in vein axils. Pale yellow, five-petalled flowers, >1.5cm across, grow in clusters of >10 from a 10cm-long bract. Fruits are hard and round. **HABITAT** Native to North America, growing in lowland woodland, riversides and urban sites. Supports a wide range of wildlife with its leaves, nectar and fruits. **DISTRIBUTION** Occasionally planted outside its native region as an ornamental.

French Tamarisk
■ *Tamarix gallica* Height >8m

DESCRIPTION Straggly, windswept deciduous small tree or large shrub with numerous fine, brittle branches and fibrous, purplish-brown bark. Very small leaves are greenish-blue, scale-like, >2mm long and clasp young shoots. Minute pink, five-petalled flowers grow in dense, tapering racemes >2.5cm long; each petal is <2mm long. Seeds are wind dispersed. **HABITAT** Native to SW Europe. Favours coastal sites, cliffs, dunes and riverbanks, and sometimes occurs inland on sandy soils. **DISTRIBUTION** Long established in our region; planted as a windbreak and for soil stabilization, and can be invasive.

Sea-buckthorn

■ *Hippophae rhamnoides* Height >11m

DESCRIPTION Multi-stemmed, dense shrub or suckering small tree with fissured, peeling bark and thorny twigs covered with silvery scales that rub off. Leaves are >6cm long and 1cm wide, with silvery scales. Insignificant flowers, >3mm across, lack petals; they open in spring, with male and female flowers on different trees. Fruits are bright orange berries, >8mm long, often produced in abundance, and are an important food for migrant birds. **HABITAT** Native to Europe, including E England, growing in coastal sites, especially dunes and saltmarshes, and sometimes inland. **DISTRIBUTION** Widespread along coasts; planted elsewhere to stabilize dunes, and also inland for ornament.

Handkerchief Tree

■ *Davidia involucrata* Height >20m

DESCRIPTION Slender, conical deciduous tree with a stout, tapering bole and orange-brown bark, peeling vertically. Leaves are >18cm long and heart shaped, each with a pointed tip, toothed margins and 15cm-long, pinkish or yellow-green petiole. They are dark shiny green above, paler and downy below. Flowers are small and petal-less, and grow in dense clusters of many male flowers with purple anthers and one hermaphrodite flower; surrounded by large pair of white bracts, one larger than the other, >20cm long. Rounded fruits, >2.5cm across, are green at first, ripening to purple-brown. **HABITAT** Native to China, where it is scarce, favouring woodland, hillsides, gorges and cliffs. **DISTRIBUTION** Popularly planted in our region as a decorative garden tree.

Dogwood

■ *Cornus sanguinea* Height >4m

DESCRIPTION Shrub, or sometimes small tree on a slender bole, with smooth grey bark. Dark red winter twigs are distinctive after leaves have fallen. Leaves are opposite, oval and pointed, each with entire margins and 3–4 pairs of prominent veins. If a leaf is snapped and the two halves are gently pulled apart, stringy latex appears where the veins were broken and connects the two halves of leaf. Leaves become a rich deep red in autumn. White flowers are small, but are grouped in large terminal clusters. Fruit is a blackish, rounded berry, borne in clusters. **HABITAT** Native to Europe, in hedgerows, scrub, copses and roadsides, mostly on calcareous soils. **DISTRIBUTION** Widespread across Europe; often planted along roadsides, and seeds are dispersed by birds.

Cornelian-cherry

■ *Cornus mas* Height >8m

DESCRIPTION Small, spreading deciduous tree with an untidy crown, reddish-brown bark and mostly level branches that end in numerous greenish-yellow, slightly downy twigs. Leaves are opposite, short stalked, ovate and pointed, >10cm long and 4cm wide with rounded bases. They are dull green above and slightly downy below with entire margins. Flowers grow in small, stalked heads of >25 small yellow flowers, each about 4mm across, opening early in the year before leaves. Fruit is a short-stalked, pendulous, bright red, fleshy berry, >2cm long, with a pitted apex and acid taste. **HABITAT** Native to central and SE Europe, favouring scrub, open woodland, hillsides and cliffs. **DISTRIBUTION** Grown outside its native region for its winter flowers and edible fruits; occasionally naturalized.

Rhododendron ▪ *Rhododendron ponticum* Height >5m

DESCRIPTION Evergreen ornamental shrub with reddish, scaly bark and dense, tangled branches. Leaves are shiny, leathery, elliptical and dark green. Flowers are 4–6cm long, bell shaped and pinkish-red; they are borne in clusters in spring to early summer, and produce a spectacular display. Fruits are dry capsules that contain numerous flat, wind-dispersed seeds. **HABITAT** Native to the Himalayas. Grows in woodland, and on hillsides, ravines and mountain slopes, favouring acid soils and wet climates. **DISTRIBUTION** Widely planted in our region as an ornamental garden subject; naturalized in some areas. Often controlled due to its invasive habits and ability to exclude native flora.

Strawberry Tree ▪ *Arbutus unedo* Height >9m

DESCRIPTION Small, spreading evergreen tree with a short bole, dense, domed crown and reddish bark peeling away in shreds. Branches are ascending and twisted; twigs are slightly hairy and reddish. Leaves are >11cm long, with either sharply toothed or entire margins, and a prominent midrib. Flowers are borne in pendulous clusters at the same time as fruits from the previous year; flowers are white, >9mm long, and sometimes tinged pink or green. Fruit is a round berry, >2cm across; its warty skin ripens to deep red. **HABITAT** Main native range is SW Europe and Mediterranean; also occurs naturally in SW Ireland in open woods and thickets. Grows in woodland and scrub, and on hillsides and cliffs, often in gardens. **DISTRIBUTION** Planted widely outside its native range.

Ash ▪ *Fraxinus excelsior* Height >40m

DESCRIPTION Large deciduous tree with a straight bole and high, open, domed crown. Bark is smooth and pale grey in young trees, but in older trees it becomes vertically fissured. Branches are mostly ascending and terminate in grey twigs tipped with sooty-black buds. Leaves are pinnate and >35cm long, each bearing 7–13 pointed, toothed leaflets. Small flowers are purple and borne in clusters near tips of twigs in spring. Male and female flowers are separate and mostly occur on separate trees. Fruits are single-winged 'keys', hanging in bunches. **HABITAT** Native to most of Europe, including Britain and Ireland. Favours woodland, limestone uplands, hedgerows, copses and urban sites, on heavy and base-rich soils. **DISTRIBUTION** Widespread throughout its range.

Manna Ash

▪ *Fraxinus ornus* Height >24m

DESCRIPTION Medium-sized deciduous tree with a flattish crown, smooth, dark grey or sometimes almost black bark, and smooth grey twigs, sometimes tinged yellow, ending in greyish, white-bloomed buds. Leaves are opposite, pinnate and >30cm long, with up to nine ovate, toothed leaflets that have white or brown hairs on veins beneath. Showy, creamy-white, fragrant flowers open with leaves. Each flower has four petals about 6mm long; flowers hang in clusters about 20cm across. Narrow-winged fruits, >2cm long, hang in dense clusters. **HABITAT** Native to central and S Europe and SW Asia. Grows on hillsides, and in woodland margins and clearings, and urban sites. **DISTRIBUTION** Planted in our region as a street tree or for ornamental value.

Wild Privet

■ *Ligustrum vulgare* Height >5m

DESCRIPTION Much-branched, semi-evergreen, spreading shrub, very rarely reaching the size of a small tree. Branches are dense and much divided, with downy young twigs. Leaves are shiny, untoothed, oval, opposite and semi-deciduous. Flowers are 4–5mm across, creamy-white, fragrant, four petalled and borne on terminal spikes. Fruits are shiny, globular and poisonous, ripening to black in autumn; they are borne in clusters. **HABITAT** Native to much of Europe, growing in scrub patches, hedgerows, and open areas on chalk and limestone; often forms thickets. **DISTRIBUTION** Widespread in suitable habitats across Europe. Frequently planted for hedging and clipped into shape, when it does not flower; often naturalized.

Lilac ■ *Syringa vulgaris* Height >7m

DESCRIPTION Small deciduous tree, and sometimes little more than a multi-stemmed shrub with a rounded crown and short bole surrounded by suckers. Greyish bark is spirally fissured in older trees. Usually a mass of ascending branches; twigs are rounded and shiny greenish-brown. Leaves are opposite, >10cm long, ovate or slightly heart shaped with entire margins; they are usually yellowish-green with a smooth surface. Fragrant flowers are borne on dense, paired, conical spikes, >20cm long, arising from leaf axils. Fruit is a pointed ovoid capsule >1cm long. **HABITAT** Native to Balkans. Favours rocky hillsides, open, sunny areas, thickets and scrub. **DISTRIBUTION** Long cultivated throughout Europe for its attractive fragrant flowers; frequently naturalized.

Indian Bean Tree ■ *Catalpa bignonioides* Height >20m

DESCRIPTION Medium-sized deciduous tree with a short bole, greyish-brown, scaly bark and spreading branches with smooth, stout twigs. Leaves are long stalked, large, broadly ovate, >25cm long, and have heart-shaped bases, short-pointed tips and downy undersides.

Flowers are 5cm long and bell shaped, with two lips; petals are white with purple and yellow spots. Flowers grow in large, showy panicles. Fruits are slender, bean-like pods, each >40cm long, and hang from branches after leaves have fallen; fruit contains many flat, papery seeds, >2.5cm long. **HABITAT** Native to SE USA. Mostly seen in cultivation, but occurs naturally in river valleys and lowland areas. **DISTRIBUTION** Planted in Europe as an ornamental and quite common in many large cities, including London.

Yellow Catalpa
■ *Catalpa ovata* Height >12m

DESCRIPTION Very similar to Indian Bean Tree (see above), and best distinguished by comparing the leaves, which are dark green and pentagonal, with a short point on each corner; they are large, >25cm in each direction, with a heart-shaped base. Branches are spreading and bark is grey-brown and scaly. Off-white flowers, >2.5cm across, are tinged yellow and spotted red inside; they grow in showy spikes about 25cm long. Pod is about 25cm long. **HABITAT** Native to China. Grows in woodland margins, on hillsides and on mountain slopes >2,500m. **DISTRIBUTION** Introduced to our region for ornament; rarely seen outside parks and gardens.

Elder ■ *Sambucus nigra* Height >10m

DESCRIPTION Small deciduous, rather untidy tree or large shrub; bole is normally short and often has young shoots emerging from it. Deeply grooved and furrowed bark is greyish-brown, often with a corky texture in older specimens. Branches are spreading and twisted. Compound leaves have 5–7 pairs of leaflets, each one >12cm long, ovate and pointed, with toothed margins and a slightly hairy underside. Crushed leaves have an unpleasant smell. Sickly sweet-scented flowers are borne in dense, flat-topped clusters >24cm across; individual flowers are small and composed of 3–5 white petals and anthers. Fruit is a rounded, shiny black berry, often produced in great quantities in pendulous heads. **HABITAT** Native to much of Europe except extreme north. Grows in hedgerows, woodland margins, scrub and wasteground in soils with a high nitrogen content. **DISTRIBUTION** Widespread and common.

Guelder-rose
■ *Viburnum opulus* Height >4m

DESCRIPTION Small, sometimes rather spreading deciduous tree, with reddish-brown bark; twigs are smooth, angular and greyish. Opposite leaves, >8cm long, each have 3–5 irregularly toothed lobes and thread-like stipules. They often turn a deep wine-red in autumn. White flowers are borne in flat heads, resembling those of a hydrangea: large, showy outer flowers are sterile, and smaller flowers in the centre are fertile. Fruit is a rounded, glistening, translucent red berry that hangs in clusters on tree after leaves have fallen. **HABITAT** Native to much of Europe, including Britain and Ireland, growing in damp woodland, hedgerows, thickets and roadsides; favours calcareous and neutral soils. **DISTRIBUTION** Widespread in suitable sites.

Wayfaring-tree
▪ *Viburnum lantana* Height >6m

DESCRIPTION Small, spreading deciduous tree with dark brown bark and rounded, greyish hairy twigs; through a hand lens the hairs look star shaped. Leaves are opposite, >14cm long, ovate, rough to the touch and have toothed margins; the undersides are thickly hairy, with more stellate hairs. Flowers are produced in rounded flowerheads about 10cm across, comprising numerous white fertile flowers; each one is about 8mm across with five white petals. Fruits are flattened oval berries about 8mm long, red at first but ripening to black; ripening is staggered, giving a striking mixture of red and black berries side by side. HABITAT Native to most of Europe except extreme north. Grows in woodland margins, hedgerows and scrub patches, preferring dry, chalky soils. DISTRIBUTION Common and widespread in suitable sites, and sometimes planted along roadsides.

Laurustinus
▪ *Viburnum tinus* Height >7m

DESCRIPTION Small evergreen tree with attractive glossy foliage and flowers produced freely in winter. Branches bear faintly angled, slightly hairy twigs. Leaves are opposite, >10cm long and oval, with entire margins, dark green and glossy upper surfaces, and paler, slightly hairy lower surfaces. Pink and white flowers are borne in branched, rounded clusters, >9cm across; individual flowers are about 8mm across with five petals, and pink outside and white inside. Rounded fruits are about 8mm long and steel-blue when ripe. HABITAT Native to Mediterranean region. Favours open, sunny hillsides, scrub and woodland margins. DISTRIBUTION Hardy and thus widely planted outside its native region as a garden shrub or tree, and also used for hedging and shelter; occasionally naturalized.

Cabbage Palm ■ *Cordyline australis* Height >13m

DESCRIPTION Superficially palm-like evergreen. Trees that have flowered have a forked trunk with a crown of foliage on top of each fork. Bark is pale brownish-grey, ridged and furrowed. Tall, bare trunks are crowned with dense masses of long, spear-like, parallel-veined leaves, >90cm long and 8cm wide. Flowers are produced in large spikes, >1.2m long, comprising numerous small, fragrant, creamy-white flowers. Fruit is a small, rounded, bluish-white berry about 6mm across, containing several black seeds. **HABITAT** Native to New Zealand. Mainly seen in coastal sites in milder areas. **DISTRIBUTION** Planted in our region for ornament. Survives quite far north, as long as there is some protection from severe cold, and tolerates a range of soil types. Often used to create the illusion of subtropical conditions in coastal resorts.

Chusan Palm ■ *Trachycarpus fortunei* Height >14m

DESCRIPTION Palm whose tall bole is covered with persistent fibrous leaf bases that hide the bole itself. Leaves are palmate, >1m in diameter and split almost to the base; segments are stiff and pointed, usually bluish-green on the underside and dark green above. Petioles are >50cm long and toothed on the margins, with the base hidden by dense brown fibres. Fragrant yellow flowers, each with six segments, are borne on a long, branched spike, with males and females occurring on different trees. Large numbers of 2cm-long, purple-tinged fruits are produced in late summer. **HABITAT** Native to China. Favours wooded hillsides, scrub and riversides, and often planted in towns and cities. **DISTRIBUTION** Introduced into Europe as an ornamental, and common in parks and gardens. One of the hardiest palms.

USEFUL WEBSITES

The Forestry Commission: www.forestry.gov.uk
Information about forests and access arrangements, visitor centres, tree planting, woodland creation and management.

The Tree Council: www.treecouncil.org.uk
The UK's leading charity for trees.

The National Trust: www.nationaltrust.org.uk
The National Trust owns many woods and forests which can be visited, as well as gardens which are home to fine trees and shrubs.

The Wildlife Trusts: www.wildlifetrusts.org.uk
There are 47 Wildlife Trusts which cover the UK and Isle of Man and Alderney, mostly organized on a county basis, and they manage many nature reserves, some of which are woodlands which can be visited.

The Field Studies Council: www.field-studies-coucil.org
The FSC runs courses on tree identification and produces many useful publications, including easy-to-use charts.

SUGGESTED READING

Sterry, P. 2008. *Collins Complete Guide to British Trees*. HarperCollins.

May, A. and Panter, J. 2000. *AIDGAP Guides: A guide to the identification of deciduous broad-leaved trees and shrubs in winter*. Field Studies Council.

Price, D and Bersweden, L. 2013. *Winter trees: a photographic guide to common trees and shrubs*. Field Studies Council.

157